A Practical Guide to Microbiologically Influenced Corrosion (MIC) Management in the Upstream Oil and Gas Sector

Ali Morshed

AMPP™

AMPP
Association for Materials Protection and Performance

Printed in the United States of America

ISBN: 978-1-57590-424-5

AMPP
Association for Materials Protection and Performance
15835 Park Ten Place
Houston, TX 77084
ampp.org

Dedication

This book is dedicated to the Iranian women, especially the younger generation, who are the beacon of bravery, courage, determination, inspiration, hope and progress for not only Iran, but for the whole world. They deserve the best.

اندیشه نیک
GOOD THOUGHTS
گفتار نیک
GOOD WORDS
کردار نیک
GOOD DEEDS

Contents

Acknowledgments

I would like to express my deepest gratitude and appreciation to two of my best, nicest and greatest corrosion mentors: (the late) Dr. Iain Spark and Dr. Robert Howard, both from the UK. They taught me the basics and fundamentals of MIC in the early 2000s at the beginning of my professional career as a corrosion engineer.

Simultaneously, I am very grateful to Ms. Jean Broge (AMPP Product Development Manager) for all her kind, honest, and wholehearted support throughout the book publication process: from the proposal stage, to book compilation, to book promotion, to book edition, and finally to its printing. Jean also very kindly and considerately let me work again with Ms. Elaine R. Firestone as my book editor. As a lead book editor, Elaine possesses the Midas touch and transforms soulless and (sometimes) bland technical scripts into interesting, live, and easy-to-read and understand books. This has been the third time I have had the pleasure of working with Jean and Elaine and cannot thank them enough for all their help and support during the past few years.

During the early stages of writing the manuscript of this book, I received invaluable advice on content, structure, and various useful sources (e.g., books and articles) from two of the most distinguished figures in the MIC field–Mr. Richard (Rick) B. Eckert from the US, and Dr. Torben L. Skovhus from Denmark. It is a great honor for me personally (because I am not a microbiologist by education) that my initial book proposal was reviewed and approved by these two MIC gurus. I remain indebted to them indefinitely and infinitely.

I would also like to thank all those wonderful individuals–my former colleagues, peers, co-workers, employers, and clients–who provided me with great opportunities to work on MIC cases and projects, which enabled me to continually enhance my MIC-related knowledge and experience in different parts of the world such as the UK's North Sea, the Persian Gulf Region, and North Africa. I am most grateful to all of them.

About the Author

Ali Morshed received his M.Sc. degree from Imperial College London in 1997, and his Ph.D. from University College London in 2002–both in corrosion engineering. He then moved to Aberdeen, Scotland and worked for various UK companies between 2002 and 2013 as a corrosion engineer. In 2013, he joined Saudi Aramco for two years before becoming a consulting corrosion engineer in the latter half of 2015.

He has so far worked on corrosion engineering and corrosion management projects both with engineering and operating companies in the United Kingdom's North Sea, North Africa, the Persian Gulf Region, and South Asia.

Dr. Morshed is the author of the Association for Materials Protection and Performance (AMPP, formerly NACE International) book *An Introduction to Asset Corrosion Management in the Oil and Gas Industry* (2012), whose second edition was published in 2016, and the third edition in 2020. He has also authored two more books on this subject for both the hydrocarbon and the non-hydrocarbon industries. The first one was published in July 2017 and the latter in May 2018.

In addition, he has been providing corrosion management training for the hydrocarbon industry since 2005, and for the non-hydrocarbon industry since 2016. Dr. Morshed has also taught the same topic to postgraduate students at Robert Gordon University in the United Kingdom.

He is a regular contributor to AMPP's *MP* Magazine on the subjects of corrosion management and MIC. Currently, he spends the majority of his time providing corrosion management services (i.e., training, auditing, and consultancy) for both the hydrocarbon and non-hydrocarbon industries.

This book is his first book on the subject of MIC.

Preface

The main objective for this book was to produce a *practical* MIC guide to further facilitate the identification, monitoring, assessment, and control of bacterial activity and MIC that might arise within assets associated with the upstream oil and gas sector. "Practical" refers to content that can be readily related to various bacterial and MIC scenarios, which are likely to arise for any upstream operator and (the content which) can be easily used to resolve, rectify, or improve such (MIC) scenarios. This book has been written to further help and facilitate bacterial and MIC troubleshooting for the operators and asset owners concerned.

To the best of the author's ability, only content and material that have already been satisfactorily tried and tested in the field by various operators have been included. In other words, no theoretical suggestions or instructions, which are based on pure lab work or research–but were never tested in the field–are included.

To further facilitate reading this book and implementing its content, included are numerous bullet points and lists pertaining to various components, categories, important variables, or controlling parameters associated with a particular topic, process, procedure, or activity. Therefore, when discussing a new topic, process, procedure, or activity the reader can easily see and comprehend the associated components, categories, or controlling parameters.

The first two chapters of this book contain introductory and general material related to asset integrity management, corrosion, and MIC in the upstream oil and gas sector.

Chapter 3 provides a detailed account on bacterial and MIC fundamentals including bacterial growth conditions and nutrition. Understanding this chapter significantly helps the asset operator or owner to better control and mitigate bacterial activity and MIC.

Chapters 4, 5, and 6 cover topics associated with bacterial and MIC monitoring, assessment, and control, respectively. These three topics are the three stages that compose the overall bacterial and MIC mitigation process. Comprehending each of these three chapters is a prerequisite for creating and maintaining an efficient bacterial and MIC mitigation process for any upstream asset.

Chapter 7 focuses on shortcomings associated with each stage of the above process. It is strongly recommended for any operator who implements this process to thoroughly read this chapter to ensure they are not making similar (operational) mistakes leading to similar MIC shortcomings.

Chapter 8 contains numerous recommendations that are useful for any particular upstream operator who is willing to enhance the efficiency of their existing bacterial and MIC mitigation process.

Appendix A contains several case studies from all over the world. Reading through such case studies demonstrates how (sometimes) very trivial issues and oversights led to acute, costly, and severe bacterial contamination and MIC issues months or years later. They also show how easily such MIC issues could have been avoided in the first place, provided that adequate (bacterial and MIC) competency existed among the personnel and managers concerned.

Lastly, Appendix B is a list of the abbreviations and acronyms used in this book and in the corrosion literature.

1 Introduction

1.1 Background

The upstream oil and gas assets are exposed to numerous integrity threats, one of which is corrosion. Corrosion in such assets often manifests itself through various deterioration mechanisms, among which microbiologically influenced corrosion (MIC) is one of the most insidious ones.

MIC is not only detrimental to the upstream assets, but if not adequately controlled it has the potential to damage the associated production reservoirs through souring and pore plugging mechanisms. MIC is a highly localized corrosion deterioration mechanism often associated with the lower half (or more precisely the "6 o'clock position") of the affected equipment.

MIC can virtually affect any process and utility system that contains even traces of water. Such systems include

- Seawater
- Fire water
- Cooling water
- Produced water
- Production systems (i.e., wet oil and gas systems)
- Drains
- Product storage (e.g., tanks)
- Product transfer (e.g., pipelines)

Because of its inherent high deterioration rates, MIC can cause leaks in a relatively short period of time and can inflict additional repair and maintenance costs on asset owners and operators. Production shutdowns, due to MIC leaks, can become a common and frequent occurrence, inducing further costs. Thus, it is highly sensible and desirable to create and maintain an MIC mitigation plan or policy for the whole duration of the asset life cycle: from design through fabrication, transportation, installation, and hydrotesting, and finally to the operations and maintenance phase.

The bacterial and MIC mitigation process itself is composed of the following three stages with the pertinent activities listed under each stage:

1. The bacterial and MIC monitoring stage
   a. Fluid and biofilm[1] sampling
   b. Corrosion rate monitoring (e.g., corrosion weight loss coupons and probes)
   c. Wall thickness inspections
   d. Pipeline intelligent pigging
2. The bacterial and MIC assessment stage
   a. Traditional or culture-based techniques
   b. The new molecular microbiological methods (MMM)
   c. Analysis, trending and processing the sampling, inspection, corrosion rate monitoring, and weight loss data)
3. The bacterial and MIC control stage
   a. Water removal (or product dehydration)
   b. Chlorination
   c. Biociding
   d. Nitrate treatment
   e. Pipeline (cleaning) pigging
   f. Corrosion management applications

To establish and implement each of these three stages, one has to be fully conversant with the relevant basics and principles associated with bacteria and MIC. Learning such basics and their timely and rigorous implementation will markedly improve bacterial and MIC mitigation throughout an asset life cycle. Simultaneously, effective and successful bacterial and MIC mitigation will lead to optimized corrosion and integrity management costs. An effective bacterial and MIC mitigation plan will also enhance an operator's MIC failure preemption capability during the asset operation phase, leading to fewer numbers of production shutdowns.

In general, MIC has proven to be a tricky corrosion deterioration mechanism for many upstream asset operators, causing frequent operation shutdowns and production reduction rates. However, adequate competency, in relevant areas (with regards to bacteria and MIC) can significantly improve the efficiency of any MIC mitigation plan or policy leading to

- Enhanced site personnel safety
- Increased environmental protection
- Extended asset operational life
- Reduced integrity and leaks issues (due to MIC)

1. See "Section 1.6 Glossary" for definition.

1.2 Rationale

In the 20 years that the author has been working as a corrosion engineer, MIC-related cases, issues, and leaks associated with both hydrocarbon and non-hydrocarbon industries have been frequent occurrences. These MIC cases were conspicuously different from each other in terms of

- Severity (i.e., deterioration rate)
- Location (i.e., process and utility systems affected)
- Cost of the associated repairs and replacements
- Extent of the production shutdown and the associated deferred production
- Consequence(s) of equipment failure
- Extent of bacterial contamination (i.e., types and populations)

Simultaneously, nearly all such MIC cases had the following attributes in common:

- The root causes were the total lack of, or inadequate (MIC-related), competency among the relevant personnel. The observed incompetency was related to basic bacterial and MIC fundamentals or principles such as
  - Bacterial growth conditions and nutrition requirements
  - Liquid and biofilm sampling
  - Equipment and process design
  - MIC assessment activities
  - MIC control activities (e.g., water removal, chlorination, and biociding)

The majority of the observed cases could have been easily and efficiently controlled and mitigated initially, had the relevant personnel received or possessed adequate (bacterial and MIC-related) training and experience, respectively.

Significant equipment repair and replacement costs could have been avoided provided that adequate (MIC) competency existed among the pertinent personnel.

In summary, based on the author's personal field experience, MIC-related incompetency has been the main culprit behind the majority of the encountered cases, followed by the lack of appropriate corrosion management applications. Simultaneously, the author does not agree with some colleagues and fellow corrosion engineers who generally attribute a considerable portion of the encountered MIC issues (in the upstream oil and gas sector) to the absence of, or the inadequate use of, new assessment technologies and methodologies (such as molecular microbiological methods [MMM]). Based on such notions, using new MIC assessment techniques will further mitigate bacterial contamination and MIC leaks across upstream assets. However, it should be carefully noted that using new assessment technologies does not improve or rectify the much-needed MIC-related competency among the pertaining personnel.

As a corrosion engineer, the author has worked with operating companies who had unlimited access to the latest bacterial assessment technologies and yet they failed miserably (in terms of the number of MIC leaks identified and the severity and extent of bacterial contamination within their various assets). The main observed root cause was again inadequate MIC-related competency among the relevant site personnel. Conversely, the author has also had the pleasure of working with some operating companies who (for various reasons) did not use or did not have access to the newer MIC assessment techniques and yet they were very successful and effective in MIC control and mitigation across their assets. The author can honestly attribute such success and effectiveness to the adequate bacterial and MIC competency (through continuous and proactive training programs) among their relevant site personnel.

While the author does not dismiss the advantages offered by the new MIC assessment technologies, nor does he see them as the silver bullet in effective and successful MIC mitigation when the pertinent personnel are not fully conversant with the aforementioned principles associated with bacterial and MIC basics. The overall bacterial and MIC mitigation process is composed of the following three stages:

1. The bacterial and MIC monitoring stage (e.g., fluid and biofilm sampling, wall thickness inspections, pipeline intelligent pigging, and corrosion rate monitoring)
2. The bacterial and MIC assessment stage (e.g., bacterial identification and enumeration techniques)
3. The bacterial and MIC control stage (e.g., water removal, chlorination, and biociding)

It is clear that MIC assessment is only one part or component of the overall bacterial and MIC mitigation process–one should not expect effective MIC mitigation through using proper MIC assessment techniques while the other two stages have not been carried out adequately, competently, effectively, and in a regular and timely manner.

Further supporting this argument is the following case associated with an upstream operating company. This case demonstrates how shear lack of (MIC-related) competency, along with a strong dose of ignorance, detrimentally affected an asset's integrity.

The case pertains to a huge seawater treatment plant, which provided treated (i.e., deoxygenated, chlorinated, and biocide-treated) seawater to maintain reservoir pressure through a pipeline network with a total length of more than 500 km and the associated injection wells.

A newly appointed plant manager (who was also very keen on making huge integrity management cost savings) ordered biocide treatments be halted permanently because he could not see the threat–that is the bacteria–with his naked eyes. In his opinion, if a threat was not visible (with naked eyes) then it was not worth the money to mitigate it! In spite of repeated appeals from his corrosion and inspection engineers, he upheld his decision for a couple of years until he was fired from the company because of spiraling MIC-related leaks within the plant and the downstream pipeline network, the increasing operations shutdowns, and the associated huge repair costs.

This scenario had happened several years ago—unbelievable in the 21st century—and within an upstream production company with unlimited access to the latest available MIC assessment technology. However, no amount of state-of-the-art technology can take the place of having virtually no competency pertaining to basic bacterial and MIC principles among its relevant site personnel and managers.

Another rather common and upsetting case, which is again a direct consequence of inadequate MIC-related competency, is the purchase of biocide chemicals by various operating companies that are only capable of killing the "planktonic" bacteria but do not have any effects on the "sessile" bacteria[(2)] or biofilms. When the author conducted reviews to find out why such chemicals were being purchased, he realized that the responsible personnel did not know that apart from the floating bacteria (i.e., the planktonic ones)—which were easier to kill—there was another type (i.e., the sessile ones), which could not be killed as easily as the planktonic ones. In some cases, the operating companies were misled by the chemical suppliers, but nobody in the operating companies concerned had requested lab test results or certificates showing the purchased chemical's ability to kill and destroy sessile bacteria or biofilm. This is because they knew almost nothing about bacterial types and the existing differences between planktonic and sessile bacteria.

Based on such personal experience, the lack of MIC-related competency and experience has been the main culprit behind the majority of the encountered MIC cases. Therefore, this book has been compiled with the intention to address such (MIC-related) competency issues as much as possible.

1.3 Main Objectives

As was mentioned in Section 1.1 (Introduction) the bacterial and MIC mitigation process is composed of three stages.

For a bacterial and MIC mitigation process to be efficient and successful, all three stages must be designed, planned, and implemented not just properly, but also competently. In turn, proper planning and implementation of each of these three stages depends very much on adequate MIC-related competency among the pertinent personnel. More precisely, this MIC-related competency pertains to those items in the list that are preceded by lowercase letters (see Section 1.1). These items, called "activities," are associated with any of the three stages of the bacterial and MIC mitigation process. This means, when these activities are planned and implemented more competently, the results show an improved and more efficient implementation of the overall MIC mitigation process. Consequently, the main objectives of this book can be defined as in the following:

2. For the definitions of these bacteria types, see "Section 1.6 Glossary."

1. To provide practical guidance associated with the listed activities. These activities control and influence the performance of the three main stages that comprise the overall bacterial and MIC mitigation process. The provided practical guidance covers the following topics:
    - Bacterial growth conditions and nutrition fundamentals
    - Fluid and biofilm sampling
    - Laboratory testing and assessment
    - Water removal or product dehydration
    - Chlorination and biocide treatment
    - Equipment design
    - Inspection
    - Corrosion rate monitoring
2. To provide corrosion management guidelines associated with each of the three stages that comprise the bacterial and MIC mitigation process to render these stages more streamlined, efficient, and successful
3. To improve the MIC failure preemption capability across the associated assets (hence progressively reducing the number of MIC-related leaks)
4. To help to further optimize MIC-related integrity management costs
5. To share international experience, knowhow, and best practices

1.4 Scope

This book has been compiled specifically for the upstream oil and gas sector covering the following asset types:

- Drilling facilities
- Offshore and onshore exploration, production, and processing plants
- Product storage facilities
- Water injection systems (in particular)
- Utility systems (that contain water)
- Associated flowlines and pipelines

The majority of both bacterial and MIC basics and principles (covered in Chapters 3–6) are identical, valid, and applicable across other oil and gas sectors, as well as other industries. Therefore, the majority of the book's guidelines and recommendations can similarly and effectively be used and applied not only in the midstream and downstream oil and gas sectors, but also in other non-hydrocarbon industries where MIC is also an integrity threat. Thus, the book content is also relevant and applicable for the following asset types.

- Midstream oil and gas sector: all product transfer pipelines–either subsea, onshore, buried, or exposed–and transfer water pipelines
- Downstream oil and gas sector: all oil and gas refineries, product storage facilities, and petrochemical complexes and all their associated pipelines
- Non-hydrocarbon industries: water intake facilities, water and wastewater management plants, product storage facilities, product transfer pipelines, ma-

rine assets (e.g., ships and submarines), cooling water systems, and any other industry where MIC is major integrity issue

1.5 Intended Audience

This book has been compiled for anyone interested and involved in the design and implementation of the bacterial and MIC mitigation process. There are no technical prerequisites for using this book so anyone with a basic appreciation of corrosion and asset integrity management subjects can take advantage of the information and guidance herein. The following personnel are deemed to benefit the most from this book:

- Corrosion and integrity engineers and managers
- Inspection engineers
- Operation engineers and managers
- Maintenance engineers
- Asset project engineers and managers
- Cathodic protection and coating engineers and technicians
- Technical assistants
- Production chemists and chemical treatment specialists
- Process engineers
- Health, safety, and environment (HS&E) engineers
- Pipeline integrity engineers
- Corrosion monitoring specialists
- Microbiology lab technicians and engineers
- Anyone in a discipline concerned with asset integrity management, inspection, maintenance, operation, and safety

1.6 Glossary

Acid Producing Bacteria (APB): Bacteria that produce acids as an end product of their metabolism. Most acid-producing bacteria produce organic acids and a few types of bacteria also produce mineral acids.[1]

Aerobic Bacteria: Bacteria that require oxygen for growth and reproduction.[1]

Anaerobic Bacteria: Bacteria that do not grow and/or reproduce when oxygen is present.

Archaea: Single-celled microorganisms that are different from bacteria.[1]

Asset Integrity Management System (AIMS): AIMS refers to a combination of corrosion engineering-based and non-corrosion engineering-based parameters that are created and implemented to maintain and improve an asset's integrity.[2]

Bacteria: Microorganisms enclosed by a cell membrane that lack a fully differentiated nucleus.[1]

Biocide: A chemical product that is used to kill biological organisms.[1]

Biofilm: A matrix or collection of bacteria (live and dead), organic matter, debris and particulates that is adhered to a surface. [3]

Corrosion Engineering (CE): The application of design, materials selection, and environmental control principles to prevent or mitigate corrosion.[2]

Corrosion Management (CM): The process of reviewing the required integrity management measures, the regular monitoring of their performance, and the assessing of their effectiveness during the asset operation phase.[2]

DAPI: A stain–4′,6-diamidino-2-phenylindole–used in fluorescent microscopy that targets DNA in all microbial cells (living, inactive, and dead); thus, allowing visualization and enumeration of bacterial cells in a sample.[1]

Facultative Anaerobic Bacteria: Bacteria that can grow and reproduce in the presence or absence of oxygen.[1]

FISH: Fluorescence in situ hybridization is a molecular microbiological method based on gene probes targeting ribosomal ribonucleic acid (RNA) in microbial cells.[1]

Integrity Management Measure: Any parameter or variable that contributes to an AIMS and influences its functioning, effectiveness, and outcome. Examples include design, inspection, corrosion rate monitoring, fluid sampling, communication, competency, and databases.[2]

General Heterotrophic Bacteria (GHB): Bacteria that use various organic compounds (rather than carbon dioxide) as their source of carbon and energy.[1]

Hydrogenase: An enzyme that catalyzes the oxidation of hydrogen and might be involved in cathodic depolarization by sulfate-reducing bacteria.[1]

Iron Bacteria: The bacteria that oxidize ferrous ion to ferric ions.[1]

Methanogens: Microorganisms that produce methane in anoxic conditions. They are categorized as archaea.[1]

Microorganisms (MOs): Single-celled organisms of the plant or animal kingdom that range from 0.2 to 200 μm in size.[1]

Microbiologically Influenced Corrosion (MIC): A biotic corrosion process affected by the presence, activity, or both of microorganisms.[1]

Planktonic Bacteria: Bacteria that freely float in a liquid phase.[1]

Quantitative Polymerase Chain Reaction (qPCR): A molecular microbiological method used to quantify the total number of microorganisms in liquid or solid samples.[1]

Sessile Bacteria: Bacteria that adhere to surfaces.[1]

Sulfur-Oxidizing Organisms: A group of bacteria that derive energy from the oxidation of sulfur, or sulfide to sulfate. Sulfur-oxidizing bacteria are aerobic.[1]

Sulfate-Reducing Archaea: A class of archaea that reduce sulfate resulting in sulfide formation. Sulfate-reducing archaea are anaerobic.[1]

Sulfate-Reducing Bacteria (SRB): A class of bacteria that reduce sulfate resulting in sulfide formation. Sulfate-reducing bacteria are anaerobic.[1]

Integrity, Corrosion, and MIC in the Upstream Oil and Gas Industry

2.1 Introduction

Managing and maintaining asset integrity is of paramount importance for upstream asset owners and operators. The corrosion phenomenon remains as one of the main integrity threats associated with such assets.

Corrosion manifests itself through myriad deterioration mechanisms among which MIC is a prevalent one. MIC has been one of the most insidious and detrimental corrosion mechanisms encountered in the upstream sector. It not only damages the assets and their various equipment, but also has the potential to damage the associated reservoirs. Furthermore, MIC is a highly localized corrosion deterioration mechanism, often associated with severe corrosion rates.

The above traits certainly distinguish MIC among other corrosion deterioration mechanisms also encountered in the upstream sector. In spite of these malign characteristics, combating MIC remains rather straightforward and viable, provided that the pertinent personnel are fully conversant with the relevant bacterial and MIC basics and principles. Unfortunately, this had not been the case for the majority of assets visited by the author. Thus, for such assets MIC remains a tricky, tedious, and troublesome integrity threat, which often significantly increases the relevant asset integrity management costs.

2.2 Integrity and Integrity Management Concepts

In general use, the word integrity means "completeness" or "oneness." In an industrial or technical context, however, it means the equipment's ability to function in a safe manner while remaining mechanically fit. In the oil and gas industry, having integrity enables a piece of equipment to function and operate properly so that, for example, the equipment's (hydrocarbon) fluids are kept inside at all times.

Consequently, the term "integrity management" can be regarded as a continuous process throughout the asset life cycle. This process ensures that the physical or mechanical condition of the asset is maintained in such a way that there are no leaks, and all the asset equipment operate and function in a safe manner per their original design specifications or requirements.

Throughout the asset life cycle, an asset owner or operator must deal with various integrity threats. Such integrity threats (Figure 2.1) vary in nature and origin, as listed below

- Natural disasters
- Poor operations
- Enemy activity
- Effects of the external environment and the internal process fluids (i.e., corrosion, erosion, and cracking)
- Miscellaneous (e.g., design and mechanical causes)

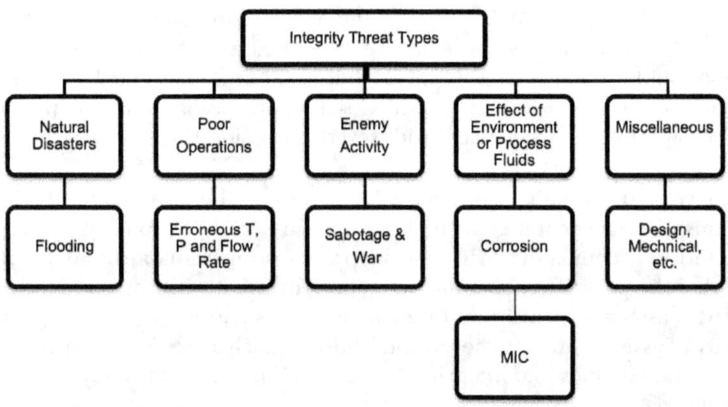

FIGURE 2.1 Integrity threat types and the associated examples.

To control, prevent, or mitigate such integrity threats for any particular asset or a collection of assets, there must exist an asset integrity management system (AIMS) where all such threats have already been taken into consideration, assessed, analyzed, and quantified. In addition, the associated preventive, mitigation, and remedial activities (against each threat) must be clearly described and devised in detail. The AIMS is a live document that should be reviewed, revised, and updated on a regular basis–where and when required.

Throughout this book, the main focus will be on the corrosion threat in general and the MIC threat in particular. Thus, the other corrosion deterioration mechanisms (such as erosion-corrosion, stress corrosion cracking, and oxygen corrosion) are not covered hereafter.

2.3 Corrosion in the Upstream Oil and Gas Sector

Corrosion poses a major integrity threat to the upstream oil and gas assets. Corrosion issues, leaks, and failures are often associated with various repercussions:

- Unplanned operation shutdowns
- Wasted product(s)
- Deferred production
- Polluted or contaminated environment
- Personnel injuries and fatalities
- Increased integrity management costs

The last item (costs) can be further broken down to the following cost categories:

- Wasted product costs (where applicable)
- Repair and replacement costs
- Deferred production costs (where applicable)
- Logistics and labor costs
- Asset downtime costs
- Legal costs and fines
- Cleaning costs (where applicable)

Therefore, proper and effective corrosion control and mitigation is very important for the upstream oil and gas assets to maintain continuous operations, and also to optimize corrosion and integrity management costs.

2.4 MIC in the Upstream Oil and Gas Sector

As already mentioned in Section 2.3, corrosion is a major integrity threat associated with the upstream oil and gas sector. The corrosion phenomenon manifests itself in the form of myriad (corrosion) deterioration mechanisms:

- General corrosion
- Pitting corrosion
- Galvanic corrosion
- Erosion-corrosion
- Environmentally-assisted cracking (EAC)
- MIC
- Oxygen corrosion
- Sweet corrosion
- Sour corrosion
- Corrosion under insulation (CUI)
- Under deposit corrosion (UDC)

Among the aforementioned deterioration mechanisms, however, MIC is not only considered as one of the most prevalent types, but it can also be regarded as one of the most distinguished, and trickiest ones because of the reasons detailed in subsequent sections (Sections 2.5 and 2.6).

MIC has so far been associated with a wide range of equipment from drilling to processing to (product transfer) pipelines. It affects both process and utility systems alike including the following:

- Wet oil and gas systems (process)
- Seawater injection systems (process)
- Produced water systems (process)
- Drains systems (process)
- Fuel systems (utility)
- Fire water systems (utility)
- Cooling water systems (utility)
- Sand wash water systems (utility)

Dead leg areas (where the flow rate is very low or stagnant conditions exist) associated with the above systems are the most ideal locations where MIC thrives and strikes the hardest.

2.5 Why Distinguish MIC Among Other Corrosion Mechanisms?

Section 2.4 listed various corrosion deterioration mechanisms often encountered in the upstream oil and gas sector, including MIC itself. However, MIC should be distinguished among those mechanisms based on a number of factors:

- MIC and bacterial control once lost are very difficult–if possible at all–to restore, particularly in more mature assets.
- It is estimated that 10–20% of all corrosion incidents in the upstream sector are due to microorganisms and MIC.[4]

- MIC corrosion products (such as iron sulfide scale) can further promote corrosion through the galvanic effect with the surrounding bare metal.
- MIC is a highly localized form of corrosion, often associated with severe corrosion rates—sometimes exceeding several millimeters per year.
- MIC can not only cause integrity issues and corrosion problems for the upstream assets, but the accompanying bacteria—if not properly and adequately controlled—can also damage the associated reservoirs through various mechanisms, leading to product souring and reduced production rates.

The above points further highlight and demonstrate the significance of creating an effective MIC mitigation plan (or process) for the upstream assets concerned, and proper and timely implementation of such a plan.

2.6 Why is Combating MIC So Complicated?

The author's international experience has shown that in most cases (associated with the assets visited) combating MIC had been rather complicated. Based on such experience, various shortcomings adversely affected MIC control and mitigation in the visited assets. Such shortcomings were either corrosion engineering-based parameters or non-corrosion engineering-based ones.

2.6.1 Observed Corrosion Engineering-Based Problems

- Design
  - Operating overdesigned equipment (overdesign in terms of size or diameter, which creates very low flow rates or almost stagnant conditions; in particular, in oversized water transfer pipelines)
  - Operating erroneous equipment geometries and shapes (for example, erroneous drainage facilities design, which can create conditions where the associated equipment cannot be fully drained and the remaining fluid [which also contains water] promotes MIC)
  - Existence of both design and operational dead legs created at the equipment design stage, which promotes MIC after commissioning
- Materials Selection
  - Using susceptible metallic alloys instead of using more MIC-resistant ones
  - Using MIC-susceptible alloys such as bare carbon steel where nonmetallic materials can be used
- Environmental Control
  - Operating assets with poor and inadequate water separation or dehydration
  - Not performing chlorination or biociding or both
  - Performing chlorination on an intermittent basis (instead of a continuous basis)
  - Application of chemical treatment using the wrong chemical (i.e., a chemical that does not kill bacteria or a chemical that actually promotes bacterial growth instead of killing them)

- Application of chemical treatment at the wrong concentration
- Application of chemical treatment in the wrong order (e.g., in seawater treatment systems the oxygen scavenger injection must always precede biocide injection and not the other way around)
- Application of chemical treatment at the wrong location

2.6.2 Observed Non-Corrosion Engineering-Based Shortcomings

- Inspections
  - Inspections carried out are not risk-based
  - Using wrong inspection techniques (i.e., not suitable for detecting MIC)
  - Not collecting and analyzing (pipeline) pig trash
- Corrosion Rate Monitoring
  - Installing the corrosion rate monitoring equipment at the wrong location
  - Installing the corrosion rate monitoring equipment at the wrong orientation
  - Not taking samples or slimes from the retrieved corrosion rate monitoring equipment, such as coupons
- Fluid Sampling and Analysis
  - Only taking fluid (planktonic) samples and not taking biofilm (sessile) samples
  - Taking fluid samples from the wrong locations
  - Taking fluid samples at the wrong time (e.g., not considering the batch biocide treatment timing when taking fluid samples); ideally, fluid samples should be taken both prior to and after batch biociding
  - Exposing taken fluid samples to oxidation and contamination
  - Erroneous sample analysis techniques or procedures
- Miscellaneous
  - Lack of or inadequate competency associated with anybody involved in the MIC mitigation process
  - Lack of or inadequate communication between operation, process, corrosion, inspection and maintenance teams, or their representatives
  - Lack of or inadequate use of corrosion control matrices (or integrity operating windows)
  - Lack of or inadequate use of corrosion key performance indicators
  - Lack of or inadequate use of various data, data management, and databases

The above shortcomings or causes are fully discussed in Chapter 7.

2.7 Summary

Integrity and its management are two key concepts in the upstream oil and gas sector. Simultaneously, corrosion is regarded as one of the main integrity threats for this sector's assets. Various corrosion deterioration mechanisms are encountered in the upstream sector–including MIC. However, MIC should be distinguished among other deterioration mechanisms because of its more detrimental, localized, and insidious nature, and characteristics. Such characteristics include very severe deterioration rates plus the potential to damage the associated reservoirs.

Combating MIC, as well as properly mitigating it, in the upstream sector has proven rather complicated in most of the assets visited. This was attributed to a multitude of often inherent shortcomings that were either corrosion engineering-based, or non-corrosion engineering-based. A successful and effective MIC mitigation plan (or process) must take both categories into consideration.

Bacterial and MIC Fundamentals

3.1 Introduction

Microbiology is one of the many subdisciplines of biology; it is the study of living organisms that are too small to be seen with a naked eye. Such tiny organisms are called "microorganisms" and bacteria are the most predominant type of microorganisms encountered in the upstream oil and gas sector.

Bacteria can initiate, sustain, or promote the metallic corrosion process in numerous ways. They cause localized corrosion mostly at the bottom of equipment (i.e., at or around the 6 o'clock position) with often severe corrosion rates.

Bacteria themselves are further divided into various types, among which are the sulfate-reducing bacteria (SRB)–the most prevalent and insidious ones in the upstream sector.

To grow and reproduce, bacteria need various nutrients, the most important of which are water, carbon, nitrogen, sulfur, and oxygen (or sometimes no oxygen at all). Apart from nutrition, bacteria also require suitable growth conditions such as stagnant flow conditions and favorable temperature, as well as pH.

All the chemical reactions that take place within a bacterial cell are referred to as "bacterial metabolism." Bacteria need to generate energy for survival; they achieve this through oxidizing their nutrients. If they use oxygen (as the oxidizing agent) during this oxidation process it is called aerobic (bacterial) metabolism; if other chemical species (such as sulfates) are used instead of oxygen, then it is called anaerobic (bacterial) metabolism.

Classifying bacteria provides a concise way of describing the main capabilities or chemical characteristics of the associated bacteria. Two classification methodologies are used simultaneously throughout this book. The first one is based on bacterial metabolism (thus categorizing bacteria as either aerobic or anaerobic), and the second is based on their mobility. That is, if the bacteria are freely floating in the bulk fluid, they are called "planktonic" bacteria and if they are attached into a substrate (and thus have become static) they are called "sessile" bacteria.

Certain basic MIC-related concepts (such as planktonic and sessile bacteria) will be defined multiple times throughout this book–sometimes even multiple times within a chapter. This repetition is because of the importance attached to such concepts and so that their importance is fully appreciated by the reader. Many of the MIC-related issues and failures encountered by the author in the upstream oil and gas sector were mainly due to the lack of understanding of such basic and critical concepts.

In the upstream sector, bacteria can be found in systems, areas, or locations where water is present, even if the water is in trace amounts. No water present means no bacteria or no bacterial activity. Under favorable nutrition and growth conditions, bacteria can form biofilms on surfaces. A biofilm is a living film formed on a substrate or surface and it can comprise various components such as live and dead bacteria, polymeric substances, proteins, corrosion products, minerals, and water. The formation and presence of biofilms often contributes to corrosion.

Bacteria cause corrosion–or more precisely MIC–in different ways. They can attack an existing corrosion mitigation measure (such as a coating) and thus indirectly promote corrosion, or they can increase the corrosivity of the environment (e.g., through acid production), and thus cause corrosion directly.

Bacteria, if not adequately controlled, can also damage the associated reservoirs through reservoir souring, and formation plugging mechanisms. Therefore, among all those corrosion damage mechanisms encountered in the upstream sector, MIC is deservedly distinguished because it is the only damage mechanism that can damage both the assets and their associated reservoirs. This further underlines the significance of proper and effective bacterial and MIC control and mitigation.

3.2 Biology, Microbiology, and Microorganisms

The term "biology" was conceived more than 200 years ago to cover those parts of the natural world that were capable of reproducing and maintaining themselves.[3] More precisely, biology can be defined as the science of life and living organisms and it covers the study of their structure, functioning, origin, growth, morphology, and metabolism.

The domains of biology are continuously expanding, making it without any doubt the science of the 21st century.[3] Included, but not limited to the following, are numerous subdisciplines:

- Anatomy
- Biochemistry
- Biogeography
- Biomedicine
- Biotechnology
- Botany
- Ecology
- Evolutionary biology
- Genetics
- Microbiology
- Molecular biology
- Physiology
- Virology
- Zoology

However, the focus of this book is on microbiology and its subcategory "oilfield microbiology," which deals with the microorganisms encountered in the upstream sector and their interactions with the metallic corrosion process that induces MIC.

Microbiology is the science of living organisms that are too small to be seen with a naked eye. In other words, microbiology is the study of organisms with dimensions less than 0.1 mm. The common term for such tiny living creatures is "microorganisms." Microbiology itself is composed of the following subdisciplines:

- Microorganisms' nomenclature
- Microorganisms' classification
- Bacteria
- Bacterial classification
- Bacterial metabolism
- Habitats
- Nutrition and growth conditions
- Archaea

Microorganisms are further divided into various subcategories such as fungi, algae, bacteria, archaea, and viruses, which are collectively referred to as "microbes." Microorganisms, or microbes, may exist in the unicellular (single cell) form or in the multicellular (colony of cells) form.

Microorganisms are either eukaryotic or prokaryotic. Eukaryote microorganisms are those organisms that have a discrete nucleus; whereas prokaryote microorganisms are unicellular organisms that lack a discrete nucleus or any other organelle.[3] Prokaryotic microorganisms include bacteria and archaea, while fungi and algae are classified as eukaryotic microorganisms. In the upstream oil and gas sector, it is prokaryotic microorganisms–and in particular the bacteria–that are often encountered and induce or promote biotic corrosion. Thus, they are the main theme throughout this book.

3.3 Bacteria

Bacteria are found everywhere. Although they are unicellular microorganisms, their combined mass on our planet exceeds that of all plants and animals.[3] Bacteria grow and reproduce in a wide range of environments and were the first living things on Earth.[3] Their range of form includes rods, spheres, spirals, and a few others. Bacteria exist in sizes ranging from 0.01–10 μm.

A wide range of bacterial species in water-containing upstream environments and systems have been encountered so far. Some of these bacteria are capable of growing, surviving, and reproducing under extreme environmental conditions in terms of pH, temperature, and salinity. Some bacterial species are capable of forming "spores." Spores are a very resilient form produced by bacteria as a reaction to adverse environmental conditions such as extremely low water content, inadequate nutrition, extreme temperatures, and the presence of biocidal chemicals. Therefore, an effective bacterial treatment or mitigation program (e.g., chemical treatment) should be able to kill both bacteria, and spores (if the latter also exists in the environment).

Bacteria are the most prevalent form of microorganisms encountered in the upstream oil and gas sector. They initiate, promote, or sustain the metallic corrosion process either directly or indirectly (see Section 3.6.1.1). Carbon steel is regarded as the most susceptible alloy to bacterial attack and MIC among all the different alloys encountered in the upstream sector. The morphology of the biotic corrosion process or MIC, is localized, and it often occurs at the lower half of the equipment. That is, the orientation of the MIC attack is mostly located at or around the 6 o'clock position.

Bacteria are divided into various types or classes, among which the SRB are both the most common and the most insidious bacterial species encountered. Hence, most of the following chapters and sections have been dedicated to topics related to bacteria (among other types of microorganisms) in general, and SRB in particular.

3.3.1 Bacterial Nutrition and Growth Conditions

3.3.1.1 Bacterial Nutrition

To sustain their growth and reproduction, bacteria must extract from their surroundings all the substances (or more precisely the nutrients) needed for both the synthesis of cellular material and generation of energy. The following are the main bacterial nutrients.

- Water: It comprises more than 80% of the cell weight and therefore is the main required nutrient.
- Carbon: Both inorganic and organic compounds can be used by bacteria as carbon sources. The organic nutrients usually serve two purposes. They serve as a carbon source for cell synthesis, and they can be oxidized within the bacterial cell to release energy. Often, a single organic compound can act both as a carbon and energy source.
- Nitrogen: Nitrates are the main source of nitrogen for most bacteria. However, some other bacteria derive their required nitrogen supplies from ammonium compounds.
- Sulfur: Sulfates are the main source of sulfur for most bacteria, including the sulfate-reducing bacteria (see Section 3.3.4).
- Oxygen: Some bacteria require oxygen for their growth. These bacteria are called "aerobes." Interestingly, for some other bacterial types, oxygen is toxic and in its presence such bacteria are killed, or their growth is inhibited. These bacteria are called "anaerobes" and SRB belong to this category. There is a third category, "facultative bacteria." They can grow and reproduce both in the presence and absence of oxygen.

Hydrogen is also another important nutrient, which comprises less than 10% of the weight of the bacterial cell. There are also other elements such as phosphorous, zinc, and manganese, which are required in very small quantities and known as the trace elements. Remember that there are many different bacterial types and as such, their nutrition requirements are very different. However, the main bacterial nutrients are the five aforementioned items.

3.3.1.2 Bacterial Growth Conditions

The previous section discussed the main nutrition requirements for bacterial growth and reproduction. However, apart from nutrition there are other factors necessary for bacterial growth, and survival as well.

- Low or stagnant flow conditions: Bacteria dislike any undue agitation and thrive in environments where stagnant flow conditions exist, or where the flow rate is low enough to let them settle onto the bottom of equipment. This explains why bacterial contamination and MIC manifest themselves at or around the 6 o'clock position of various equipment.
- Temperature: Bacteria thrive in a wide range of temperatures; however, as a rule, they grow and reproduce faster with increasing temperature.
- Presence of substrate: Nutrition concentration is considerably higher at or near a suitable substrate or surface compared to anywhere else in the bulk fluid. Therefore, during stagnant or low flow conditions when bacteria are able to migrate or settle onto the bottom of equipment, they have greater access to the required nutrition than when they are floating in the bulk fluid. A metallic substrate–in addition to providing a place for bacteria to settle–also supports various chemical reactions which are necessary to sustain bacterial metabolism.

- pH: Like temperature, bacteria can survive in a wide range of pH values from nearly absolute acidic to severe alkaline conditions. However, for each type of bacteria there is an ideal pH range for their optimum growth and reproduction.

3.3.2 Bacterial Metabolism

Bacteria are divided into various types or classes based on their ability to carry out certain functions, or more precisely based on the main chemical reactions they induce or the chemical species they consume or produce. The term (bacterial) metabolism refers to all the chemical reactions that take place within the bacterial cell. Such chemical reactions are indispensable for bacterial survival and facilitate the conversion of nutrients into new cell material and energy. The most important energy-generating reactions are those concerned with the oxidation of various nutrients as the sources of energy. In general, there are two types of bacterial metabolism.

- Aerobic metabolism: When the bacteria use oxygen to oxidize their nutrients. Aerobic metabolism can be generalized via the following chemical reaction:

$$\text{Carbon source(s)} + O_2 \rightarrow CO_2 + H_2O + \text{energy} \qquad \text{(R.3.1)}$$

- Anaerobic metabolism: When bacteria use chemical compounds other than oxygen, such as sulfates and nitrates to oxidize their nutrients. Anaerobic metabolism can be generalized via the following chemical reactions:

$$\text{Carbon source(s)} + SO_4^{2-} \rightarrow CO_2 + H_2S + \text{energy} \qquad \text{(R.3.2)}$$

$$\text{Carbon source(s)} + NO_3^{-} \rightarrow CO_2 + NO_2^{-} \text{ or } NH_3 \text{ or } N_2 + \text{energy} \qquad \text{(R.3.3)}$$

3.3.3 Bacterial Classification

The bacteria encountered in the upstream oil and gas sector belong to various groups or classes. It is therefore both useful and practical to categorize or classify bacteria using a criterion such as shape, metabolism, chemical functions (i.e., the principal chemical reactions associated with each bacterial type), growth conditions (e.g., temperature), or mobility.

The main purpose of bacterial classification is to provide, by using the assigned name or class, a concise way of describing the main capabilities or (chemical) characteristics of the associated bacteria. Two different classification methodologies used throughout this book are described in this section. It is strongly recommended to use them simultaneously or in a complementary manner. This way, more useful information regarding the associated bacteria is provided. Such information will in turn facilitate decision-making regarding the required sampling, monitoring, assessment, and control procedures.

The main chemical function of the bacteria (i.e., the chemical species being oxidized, reduced, or produced) is used to distinguish or classify them. Using this methodology enables one to easily group or categorize bacteria based on their nutritional or chemical species they consume or produce.

- Sulfate-reducing bacteria (SRB): The most prevalent and insidious bacteria encountered in the upstream oil and gas sector, which reduce sulfate to sulfide through an anaerobic metabolism.
- Sulfur-oxidizing bacteria (SOB): A bacterial class that oxidizes sulfide or elemental sulfur to sulfate through an aerobic metabolism.
- Iron-oxidizing bacteria (IOB): A bacterial class that oxidizes ferrous ions to ferric ions through an aerobic metabolism.
- Nitrate-reducing bacteria (NRB): A bacterial class that reduces nitrate to nitrite and ammonia.

A secondary methodology, which complements the above classifications, is based on bacterial mobility in their environment, i.e., planktonic (freely moving or floating in the liquid phase) and sessile (those that stick onto a surface, or exist within a biofilm, in a static state).

The distinction between planktonic and sessile bacteria is very important because of the following:

- Bacteria are usually introduced into a new asset, system, or equipment (often through a flowing water source or a liquid phase that also contains some water) as planktonic bacteria. Such planktonic bacteria can, under favorable flow conditions (i.e., lack of flow, agitation, or disturbance), settle down onto surfaces and become sessile bacteria.
- It is sessile bacteria that promote MIC on metallic surfaces through their attachment to such surfaces.
- Killing planktonic bacteria is significantly easier and more straightforward than killing sessile bacteria.

The second (mobility) classification methodology indicates if bacteria have been involved in the biotic corrosion process (as with sessile bacteria) at the metal surfaces. Consequently, combining the two classification methodologies (metabolism and mobility) provides useful and practical information about not only the bacterial types encountered, but also the associated chemical reactions, chemical species being consumed or produced, and the possible onset of the biotic corrosion process. The following are examples for these.

- Planktonic SRB: SRB that are still floating in the liquid phase and have not yet converted into sessile bacteria. Hence, they have not yet been involved in the biotic corrosion process at the metal surfaces.

- Sessile SRB: SRB that have already settled down onto metallic surfaces or formed biofilms and are about to begin the biotic corrosion process or have already done so, under desirable nutrition, and growth condition.

3.3.4 The Sulfate-Reducing Bacteria (SRB)

The SRB are the most prevalent and insidious bacterial type encountered in the upstream oil and gas sector. They are strict anaerobes, meaning they can only thrive and metabolize in the absence of oxygen, i.e., the presence of oxygen either kills them or inhibits their growth and reproduction. Through their anaerobic metabolism, SRB convert sulfate into hydrogen sulfide (Reaction 3.2, repeated):

$$\text{Carbon source(s)} + SO_4^{2-} \rightarrow CO_2 + H_2S + \text{energy} \qquad \text{(R.3.2)}$$

The SRB can not only damage the (metallic) equipment through the biotic corrosion process (i.e., MIC), but they can also damage the associated reservoirs via a multitude of mechanisms such as the reservoir souring mechanism, and the reservoir pore plugging mechanism (see Section 3.6.1.2).

Like other bacterial types, SRB can cause highly localized corrosion, with corrosion rates that can equal or exceed several millimeters per year. Section 3.6.1.3 lists the various deterioration mechanisms through which the SRB damage their metallic substrates.

3.3.5 Locations Where Bacteria Are Found

As explained in Section 3.3.1.1, water is one of the main nutrients necessary for bacterial survival, growth, and reproduction. Therefore, bacteria are not expected or encountered where an aqueous phase does not exist. Conversely, bacteria are expected and often widely encountered in locations (within a particular asset) where an aqueous phase already exists, even in trace concentrations, provided that the other nutrition and growth conditions simultaneously exist.

In the upstream oil and gas sector, bacteria are usually found in the following process and utility systems:

- Seawater injection system
- Produced water system
- Fire water system
- Sand wash water system
- Cooling water system
- Wet hydrocarbon (e.g., oil, gas, and condensate) systems
- Product storage tanks (e.g., diesel and fuel gas)
- Wet product transfer pipelines (e.g., wet oil and gas pipelines)
- Water transfer pipelines

Knowing the possible locations where bacteria can be encountered is of particular importance, because such knowledge can facilitate improved planning and implementation of a bacterial sampling system (see Sections 4.3.1.1 and 4.3.1.2).

3.3.6 Sources of Bacterial Contamination

Bacteria or bacterial contamination can be introduced into an already clean or a newly built asset (or parts thereof) via any of the following sources. Thereafter, if the necessary nutrition and growth conditions were present, such contamination might turn into an MIC issue over time. There are numerous sources of bacterial contamination:

- Sea, river, or well water, or any other source of inlet water
- Contaminated drilling mud or well intervention fluids (which can contaminate downhole tubing)
- Hydrotest water (which is out of spec in terms of water quality, biocide content, etc.)
- Contaminated water entering the system during construction and fabrication phases (e.g., the water entering a buried pipeline system during fabrication while the pipeline is in a trench full of surface water)
- Contaminated water entering the system during repair and maintenance campaigns when the system is opened (e.g., the surface water entering a buried pipeline system when it is shut down and opened for repair)
- Contaminated fuel transferred into a new storage system during bunkering (e.g., contaminated diesel fuel, which contains water delivered into a clean, or a new diesel fuel tank)
- Contamination entering the system via (corrosion) leaks, cracks, or ruptures
- Contaminated product received from an upstream asset or system (e.g., a contaminated feed pipeline, which can contaminate the receiving asset or facility)

3.4 Archaea

An archaeon (singular for archaea) is a single-celled prokaryotic microorganism. Like bacteria the archaea lack any nucleus or any other compartments within the cell. Originally, they were believed to be only associated with exotic environments and harsh living conditions, but it has been shown that, like bacteria, they are ubiquitous.[3] For example, methanogen microorganisms are archaea; these are involved in the anaerobic production of methane (Reaction 3.4) and promote metallic corrosion through cathodic depolarization of the associated surfaces:

$$4H_2 + CO_2 \rightarrow CH_4 + 2\,H_2O + \text{energy} \qquad \text{(R.3.4)}$$

3.5 Biofilm Basics

A biofilm can be defined as an often self-sufficient community of bacteria within which various species may interact, for example, some recycle the waste or byproducts of others.[3] Alternatively, a biofilm may be defined as a living film formed on a surface composed of a very diverse population of microorganisms, predominantly bacteria. Biofilms are often composed of living (bacterial) cells, polymeric substances, DNA, proteins, enzymes, corrosion product(s), minerals, and water.

A biofilm often starts to form when under favorable flow conditions (i.e., zero or low flow rates); the planktonic bacteria settle onto the bottom of equipment and become sessile bacteria. The material produced around sessile bacteria–the polymeric network–contributes to the integrity of biofilm and acts as a physical barrier hindering both oxidizing and nonoxidizing chemicals (see Sections 6.3 and 6.4). Biofilm polymers consume oxidizing chemicals such as chlorine and its various derivatives (e.g., hypochlorous acid [HOCl]) before they can reach and kill the bacteria entrenched within the biofilm. Because of the biofilm, killing sessile bacteria requires nonoxidizing chemicals (i.e., biocides) with concentrations many times greater than those required to mitigate or kill planktonic bacteria, which do not have a biofilm.

Formation and presence of biofilms can contribute to corrosion in the following ways:

- Physical deposition thus promoting either under deposit corrosion (UDC) or blocking corrosion inhibitor access to the surface
- Production of corrosive byproducts such as H_2S, which is one of the products of SRB metabolism
- Depolarization of the corrosion cell through the removal of corrosion products (from the surface) thus promoting further corrosion
- Promoting differential aeration cells

Biofilms often form during zero or low-flow conditions, or where intermittent operations exist leading to stagnant flow conditions, provided that nutrition and other growth conditions (see Section 3.3.1.2) simultaneously exist. Therefore, formation of biofilms within dead leg areas is very likely.

3.6 MIC Fundamentals

MIC can be defined as the (metallic) corrosion due to microorganisms' presence or activity or both.[1] Because in the upstream sector bacteria are the most prevalent type or category of microorganism encountered, the focus throughout this book will be on MIC that is induced or promoted by bacteria rather than any other type of microorganism.

Bacteria promote (metallic) corrosion both directly and indirectly as discussed in more detail in Section 3.6.1.1. They can either initiate new corrosion issues, cases, or problems, or they can exacerbate ongoing corrosion issues or cases that had been initiated or caused via other corrosion deterioration mechanisms. Such phenomena occur through one of two ways:

- Increasing the corrosivity of the environment caused by bacterial metabolism (direct MIC or bacterial involvement), or
- Detrimentally affecting the performance of an existing (corrosion) mitigation measure (indirect MIC or bacterial involvement), including bacterial deterioration or degradation of an existing coating system, or bacterial consumption of the oxygen scavenger chemical

In addition to damaging the equipment through the corrosion phenomenon, if not adequately and effectively controlled, bacteria can also damage the associated reservoirs (see Section 3.6.1.2).

For MIC to occur, the following five conditions must be present simultaneously:

- A susceptible metal or alloy
- The relevant bacterial types
- The required bacterial nutrition
- The required bacterial growth conditions (e.g., zero or low flow rates enabling the planktonic bacteria to settle or attach themselves onto metallic surfaces and thus becoming sessile bacteria)
- Absence of, inadequate, or ineffective MIC mitigation measures in place

3.6.1 Predominant Bacterial and MIC Damage Mechanisms in the Upstream Sector

Bacteria and MIC, if not efficiently controlled and mitigated, have the potential to induce severe damage on both the upstream oil and gas assets, and the associated reservoirs. Sections 3.6.1.1 and 3.6.1.2 discuss such damage mechanisms, respectively.

3.6.1.1 Asset Damage Mechanisms

Asset damage mechanisms (Figure 3.1) are divided into the two categories and several subcategories listed below:

1. Indirect asset damage mechanisms
   a. Coating damage
   b. Corrosion inhibitor consumption
   c. Oxygen scavenger consumption

2. Direct asset damage mechanisms
    a. Physical deposition
        i. Inducing UDC
        ii. Blocking corrosion inhibitor access to the (metal) surface
        iii. Creating oxygen gradients
    b. Production of corrosive products
        i. Producing hydrogen sulfide
        ii. Producing organic or inorganic acids
        iii. Producing corrosion products (such as iron sulfide) which are cathodic compared to the bare metal surface, hence promoting galvanic corrosion
    c. Depolarization
        i. Removal of adsorbed hydrogen from the surface
        ii. Removal of various corrosion products from the surface

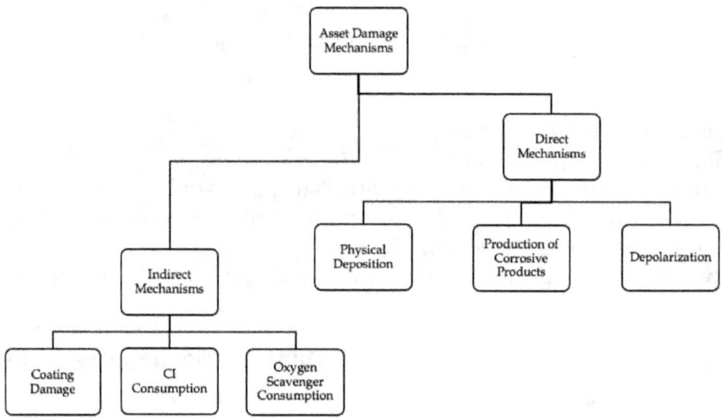

FIGURE 3.1 Asset damage mechanisms categories and subcategories.

There is a total of three indirect and eight direct damage mechanisms induced by MIC. Table 3.1 lists various corrosion damage mechanisms based on different bacterial metabolisms. The only exception in this table is the last row where the microorganism listed are archaea, not bacteria.

TABLE 3.1 Various corrosion damage mechanisms encountered in the upstream sector based on different bacterial metabolisms. The last row is an exception because the methanogens are archaea and not bacteria.

Full Name	Acronym	Metabolism	Corrosion Mechanism(s)
Sulfate-Reducing Bacteria	SRB	Anaerobic reduction of sulfate to sulfide	Refer to Section 3.6.1.3
Sulfur-Oxidizing Bacteria	SOB	Aerobic oxidation of sulfur (S^2) or elemental S to sulfate	Promotes SRB growth and producing sulfuric acid
Acid Producing Bacteria	APB	Aerobic or anaerobic production of organic acids	Lowers pH by producing various organic (and sometimes inorganic) acids
Iron-Oxidizing Bacteria	IOB	Aerobic oxidation of iron, Fe (II) to Fe (III)	Promotes SRB growth and anaerobic conditions by forming oxide layers
Iron-Reducing Bacteria	IRB	Reduction of Fe (III) to Fe (II)	Destroys surface oxide layer and cathodic depolarization
Nitrate-Reducing Bacteria	NRB	Nitrate reduction to nitrite or ammonia	Nitrite enhances pitting and nitrates are strong oxidizers
General Heterotrophic Bacteria	GHB	N/A	Promotes the attachment of other bacteria (e.g., SRB) to metal surfaces
Methanogens (Archaea)	None	Anaerobic production of methane	Cathodic depolarization

3.6.1.2 Reservoir Damage Mechanisms

What distinguishes MIC from other corrosion damage mechanisms encountered in the upstream sector is the fact that MIC, along with bacterial contamination–if not adequately and effectively controlled and mitigated–have the potential to damage the associated reservoirs as well. Such occurrence further highlights the significance of proper and effective MIC and bacterial mitigation in the upstream sector.

Below are the reservoir damage mechanisms (Figure 3.2).

- Reservoir souring–In seawater injection systems when SRB are not mitigated efficiently, the SRB can move farther downstream along the seawater transfer pipeline(s) and eventually enter and contaminate the reservoir. Provided that the required nutrition and growth conditions are present within the reservoir, their continued metabolism, growth, and reproduction will lead to higher rates of sulfate to sulfide conversion. Such increasing rates of sulfide or hydrogen sulfide production will gradually lead to greater concentrations of hydrogen sulfide within the reservoir and its hydrocarbon content; thus, the term "reservoir souring." Section 3.6.2 further discusses the repercussions of such phenomenon.

- Formation plugging–This damage mechanism is further divided into the following two subcategories:
  - Production plugging: The hydrocarbon production rate from a reservoir depends on a multitude of variables including reservoir porosity and permeability. Very often the tiny and narrow passages within the reservoir (through which the hydrocarbon fluids flow toward the production tubing) are gradually restricted and blocked (or plugged) by the growing biofilms and the production of other bacterial byproducts. This results in decreasing reservoir porosity and permeability, which in turn leads to reduced hydrocarbon production rates from the affected reservoir.
  - Injection plugging: The same mechanism described above is responsible for reducing a reservoir's ability to accept injected seawater, resulting in reduced seawater injection rates into the reservoir. This, in turn reduces production rates through the production tubing because lesser volumes of seawater can be injected into the reservoir to maintain adequate or acceptable production pressure and hence maintain acceptable production rates.

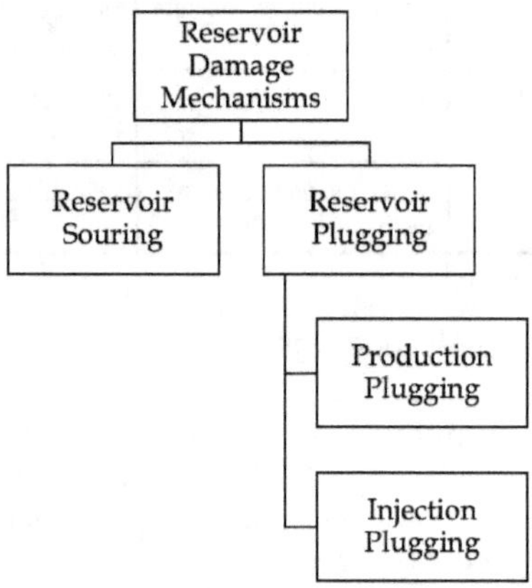

FIGURE 3.2 Reservoir damage mechanisms induced by bacteria.

3.6.1.3 SRB Damage Mechanisms

SRB–the most prevalent and insidious bacteria encountered in the upstream sector–inflicts damage on the associated equipment through the following deterioration mechanisms.

- Localized corrosion due to hydrogen sulfide production (Reaction 3.2 in Section 3.3.4).
- Localized corrosion due to sulfide production (the produced hydrogen sulfide dissociates further into hydrogen cations and sulfide anions in aqueous environments).

- Hydrogen depolarization reaction ($SO_4^{2-} + 8H_{ad} > 4H_2O + S^{2-}$)–Through this reaction (and as part of the bacterial metabolism), the SRB remove the hydrogen atoms adsorbed (represented in the reaction by the subscript "ad") on the surface, thus exposing the metal surface to the corrosive environment.
- Localized corrosion due to corrosive corrosion product–The produced hydrogen sulfide reacts with the carbon steel substrate to produce iron sulfide as the corrosion product. Iron sulfide is cathodic to the bare carbon steel areas. This creates galvanic cells where the iron sulfide scale is the cathode, the bare carbon steel areas act as the anode, and the latter corrodes via the galvanic corrosion mechanism.
- Various hydrogen damage mechanisms–The presence of sulfide in the environment (and as one of the products of SRB metabolism) retards the combination reaction of hydrogen atoms with each other (to produce hydrogen molecules [H_2]). The unreacted hydrogen atoms then spend a greater amount of time on the metal substrate, which increases the likelihood of their diffusion into the metal lattice. The diffusion of hydrogen atoms into the metal lattice (and their possible combination within the lattice to produce molecular hydrogen) promotes various hydrogen-related damage mechanisms such as hydrogen-induced cracking, hydrogen blistering, and hydrogen embrittlement.

3.6.2 Reservoir Souring Repercussions

As was discussed in Section 3.6.1.2, the inadequate mitigation of MIC and bacterial contamination within upstream assets and their equipment can lead to reservoir souring. Souring of reservoirs is associated with various repercussions for the operating companies:

- Lost revenue from the sale of gas that has gone sour
- Lost revenue from wells shut-in due to high sulfide levels
- Cost of upgrading the associated equipment to sour service specifications
- Increased corrosion and cracking failures (and thus increased number of operation shutdowns) due to sour corrosion
- Requiring the retrofit of new sweetening equipment to enable resuming gas sales or gas exports

3.7 Summary

MIC and bacterial contamination, if not properly and adequately controlled, can damage both the upstream oil and gas assets and their associated reservoirs. This further signifies the importance of having and maintaining an up-to-date and effective bacterial and MIC mitigation plan for the pertaining assets.

The science of microbiology (and in particular its subcategory of oilfield microbiology) can be extremely useful for creating, implementing, and maintaining such mitigation plans. Microbiology (or oilfield microbiology) facilitate an operator's understanding about various parameters that affect and influence bacterial life, growth and survival, as well as the locations, extent, severity, and timing associated with the MIC process. By removing, restricting, altering, or controlling some of these parameters, operators can more efficiently mitigate MIC and bacterial contamination within their assets, and their associated reservoirs.

Bacterial and MIC Monitoring

4.1. Introduction

Chapter 4 focuses on the first stage of the bacterial and MIC mitigation process, which is the bacterial and MIC monitoring stage. This stage itself is further composed of two components: bacterial monitoring and MIC monitoring. Bacterial monitoring is about taking both liquid and biofilm (or sessile) samples, while MIC monitoring is about carrying out wall thickness inspections and corrosion rate monitoring to determine or estimate the ongoing corrosion rate, which might be due to bacterial activity and MIC.

Because bacterial and MIC monitoring is the first stage of the overall bacterial and MIC mitigation process, every care should be taken to ensure that sampling, inspection, and corrosion rate monitoring data quality and accuracy is maintained and preserved. Otherwise, the assessment stage will undoubtedly produce erroneous output (in terms of bacterial types, populations, and MIC deterioration rates), which in turn can lead to wrong, irrelevant, and inefficient bacterial and MIC control (i.e., the third stage). Various parameters influence data quality during this stage including location, orientation, frequency of sampling, and inspection activities, plus the competency of those who carry out the sampling, inspection, and corrosion rate monitoring activities.

To efficiently and systematically combat and control MIC within any asset, the asset owner or operator requires a proper and thorough bacterial and MIC mitigation process to follow. Such a process is a live one and should be composed of monitoring, assessment, and control stages.

4.2 Brief Introduction to the Bacterial and MIC Mitigation Process

Throughout this book, the term "bacterial and MIC mitigation process" refers to a cyclic–and continuous–process composed of three stages (Figure. 4.1):

- Bacterial and MIC monitoring stage–The necessary sampling (both liquid and biofilm [or sessile]) along with the pertinent inspections and corrosion rate monitoring activities that are carried out (to produce the required input data for the assessment stage).
- Bacterial and MIC assessment stage–The input data produced in the first stage are evaluated, trended, processed, analyzed, and interpreted to determine bacterial types, their associated populations, and the concentration of various ions and molecules consumed or produced by the bacteria. The input data are also used to estimate or calculate the associated MIC corrosion rates.
- Bacterial and MIC control stage–In this stage, various activities are carried out to reduce the existing bacterial populations, and also to decrease the associated corrosion rates due to MIC.

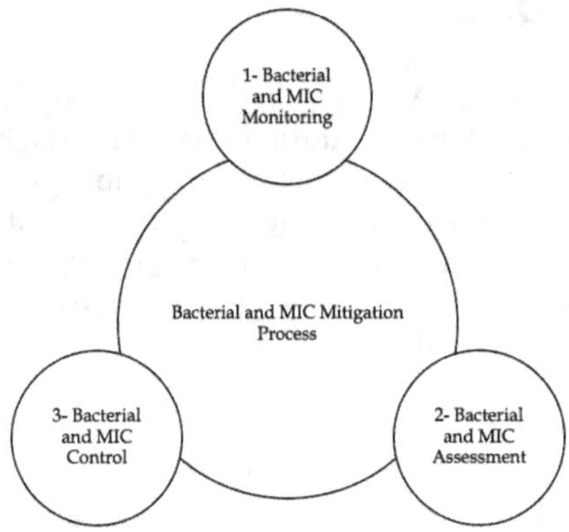

Figure 4.1 Bacterial and MIC mitigation process and its three stages.

In other words, the bacterial and MIC mitigation process is composed of three stages and each stage is composed of two components: one component pertaining to bacteria and the other to MIC. Table 4.1 provides the associated description and rationale for each of the pertinent components.

From a terminology point of view, some peers and colleagues may prefer to only use the acronym "MIC" when either referring to the aforementioned process or any of its three stages. For example, they use "MIC monitoring" instead of "bacterial and MIC monitoring," and so on. However, in the author's opinion, these two terms–bacteria and MIC–should be used together whenever referring to the above process or any of its stages because of the following reasons:

- The MIC phenomenon does not occur in the absence of bacteria or bacterial activity.
- The terms "bacteria" and "bacterial activity" are not synonymous with MIC.

Accordingly, Chapter 4 is dedicated to the bacterial and MIC monitoring stage, while Chapters 5 and 6 are dedicated to the bacterial and MIC assessment and bacterial and MIC control stages, respectively.

Table 4.1 The components associated with each stage of the bacterial and MIC mitigation process along with their pertinent rationales.

Stage	Stage Components	Rationale
Stage 1: Bacterial and MIC Monitoring	Bacterial Monitoring	To produce both liquid and biofilm (or sessile) samples for the next stage (assessment stage).
	MIC Monitoring	To produce predominantly wall thickness inspection and corrosion rate monitoring data for the next stage (assessment stage).
Stage 2: Bacterial and MIC Assessment	Bacterial Assessment	To determine types (i.e., metabolism) and populations of the bacteria encountered in the system, along with the concentration of ions and molecules consumed and produced by the bacteria.
	MIC Assessment	To determine whether or not the encountered wall losses or corrosion rates are due to MIC, and also to estimate or calculate the encountered MIC rates.
Stage 3: Bacterial and MIC Control	Bacterial Control	To use methods to either kill bacteria or retard their activity.
	MIC Control	To use methods to reduce or totally arrest the encountered corrosion rates due to MIC.

4.3. Bacterial and MIC Monitoring Stage

The main purpose for carrying out this stage is to produce the necessary input (samples and data) to be later evaluated, processed, analyzed, and interpreted during the assessment stage (Stage 2). Such input is in the form of

- Liquid samples
- Biofilm (or sessile or solid) samples
- Equipment wall thickness inspection data
- Corrosion rate monitoring data

Because bacterial and MIC monitoring is the first stage in the overall (bacterial and MIC) mitigation process, then any possible issues and shortcomings associated with the input (i.e., sampling, wall thickness, and corrosion rate data) quality and accuracy have the potential to undermine the whole mitigation process. That is, producing unreliable or erroneous input will undoubtedly result in wrong and incorrect assessment output or results. Such incorrect output will in turn induce an incorrect, inaccurate, or unrealistic picture and understanding about the size, extent, location, and type of the existing bacterial and MIC problems (in terms of bacterial types, their associated populations, and the pertinent MIC deterioration rates). This occurrence can result in wrong bacterial and MIC control decisions and erroneous remedial activities, rendering the whole mitigation process ineffective, uneconomical, and therefore useless. Thus, every care should be taken into consideration to produce highly reliable and accurate input during the monitoring stage.

The monitoring stage is composed of the following two components:

- Bacterial monitoring component
- MIC monitoring component

Sections 4.3.1 and 4.3.2 cover these two components, respectively.

4.3.1 Bacterial Monitoring Component

Bacterial monitoring produces the necessary liquid and biofilm (or sessile or solid) samples for the bacterial assessment stage, which can be carried out either in the field itself or later in a lab. The majority of the visited sites only took liquid samples; no biofilm or sessile sampling took place. However, relying solely on liquid sampling is very misleading–such an approach can produce an erroneous image (in terms of both bacterial types and populations) leading to incorrect, inadequate, or totally inefficient bacterial and MIC control measures.

It should be noted that the presence of bacteria in any system does not necessarily equate to the occurrence of corrosion–or more precisely microbial corrosion. That is why bacterial sampling must be accompanied by, and complemented with, MIC monitoring to determine whether there has been any deterioration (i.e., wall loss) in such areas and then to determine (during Stage 2) whether the observed deterioration has been due to bacterial presence, or activity, or not. Thus, accurate and effective sampling is crucial for any meaningful assessment that would take place later.

Sections 4.3.1.1 and 4.3.1.2 discuss fluid and biofilm sampling, respectively; while Section 4.3.1.3 discusses parameters that affect bacterial monitoring.

NACE TM0194 provides further instructions and guidelines on both liquid (or planktonic) and biofilm (or sessile) sampling in addition to the points highlighted and discussed in this book.[1]

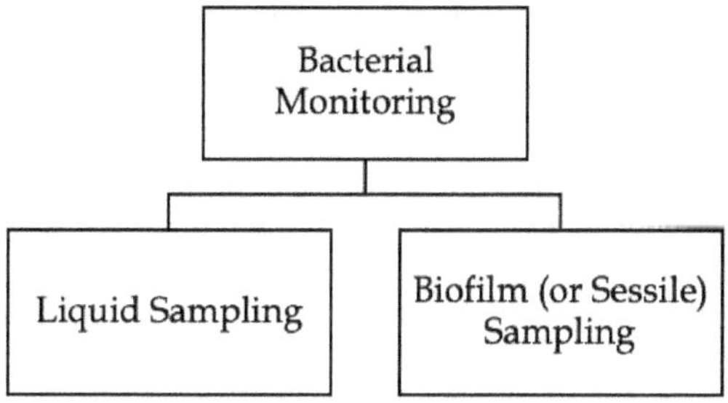

Figure 4.2 The main activities constituting bacterial monitoring.

4.3.1.1 Liquid Sampling

To determine both planktonic bacterial types and their populations in the process or utility fluids, operators need to take liquid samples from various locations on a regular basis. Liquid samples also help to determine the concentration of various ions and molecules consumed and produced by the bacteria. Such findings (i.e., the determination of species consumed and produced) can further facilitate the determination of the associated bacterial metabolism(s) and thus help the operators to identify bacterial types and populations more accurately. More accurate identification of bacterial types and their associated populations can undoubtedly result in devising better and more effective bacterial control measures and activities (Stage 3 of the overall mitigation process) leading to a more efficient bacterial and MIC mitigation process.

The operators should note that relying solely on liquid sampling (and thus only on planktonic bacterial results) will lead to significant misunderstanding about the exact location, size, extent, and rate of bacterial and MIC problems within any system. Biofilm or sessile sampling has also to take place to identify sessile bacterial types and populations.

Liquid sampling points should be selected with the utmost care to ensure that the sampling campaign covers all the critical locations (from a bacterial and MIC point of view) throughout a system or an entire asset. Apart from normal or regular liquid sampling points, one also must remember to take liquid samples from a failure or leak location—provided that safety regulations are observed. Such samples can provide valuable information about the type and population of bacteria at the failure location and the possible bacterial metabolism and deterioration mechanism, if the failure or leak was due to MIC or at least that MIC was one of the contributing factors.

Liquid samples can be taken from any of the following:

- Any available access or sample points
- Any pig trash or debris (after any pigging operation such as cleaning pigging or intelligent pigging)
- The location of a leak or failure

Apart from (later) bacterial enumerations during the assessment stage, liquid samples are taken so other relevant parameters or variables associated with water can also be measured or determined. Such parameters include

- Concentration of sulfide and sulfate ions
- Concentration of organic acids produced by bacteria
- pH
- Total dissolved solids
- Oxygen content
- Concentration of other chemical species (e.g., various cations and anions), residuals or concentrations of various antibacterial chemicals (e.g., chlorine and biocides)

Juxtaposing the information regarding the variations of the above parameters with those of bacterial types and numbers can help an operator to better appreciate the location, extent, severity, (bacterial) metabolism, and other pertinent attributes of any existing bacterial problem. Therefore, better and more efficient control measures can be devised and implemented.

4.3.1.2 Biofilm (or Sessile) Sampling

To determine both sessile bacterial types (i.e., from a metabolism point of view) and their populations, operators should take biofilm, sessile, or solid samples from various locations on a regular basis. Biofilm or sessile samples are of higher importance compared to planktonic samples, because it is the sessile bacteria that live within the biofilms and are stuck on metal surfaces that promote MIC.

Biofilm or sessile sampling was observed for a very small percentage of the sites and companies visited by the author. The majority of the visited sites only took liquid samples and no biofilm sampling was carried out. Thus, it can be said that the significance and criticality of biofilm or sessile sampling has not been fully and deservedly appreciated as yet in large portions of the oil and gas assets visited.

Biofilm or sessile samples can be taken from any of the following:

- Any removable or retractable device that provides access into the associated process or utility system

- Corrosion rate monitoring devices, such as probes and weight loss coupons, which when retracted from the system can be used to take biofilm samples
- Sidestream devices that provide easy access and quick sampling (user friendly) in a facility, along with the ability to provide long-term information on biocide efficiency on sessile bacteria
- Any pig trash or debris (after any pigging operations such as cleaning pigging or intelligent pigging)
- The location of a leak or failure

4.3.1.3 Parameters Affecting Bacterial Monitoring

The following are the parameters that affect bacterial monitoring and the quality and accuracy of its data:

- Sampling location
- Sampling point orientation (i.e., clock position of the sample point)
- Sampling frequency
- Timing regarding the chemical treatment carried out prior to or after sampling
- Taking biofilm (or sessile) samples as well as liquid samples
- Sampling procedure
- Sample storage and transport (the two most common threats are sample oxidation and contamination during storage and transport)
- Competency of the personnel planning and conducting sampling

Ignoring or not clearly defining or determining any of the above parameters can have a profound negative effect on sample quality and data reliability.

4.3.2 MIC Monitoring Component

Carrying out (liquid and solid) sampling should always be accompanied with MIC monitoring activities. While taking samples (and their later assessment) can tell us about the existing bacterial types and their associated populations, the monitoring of the MIC deterioration rate can provide us with information about ongoing corrosion rates due to bacterial activity and the pertinent remaining lives for the equipment concerned.

MIC monitoring is often achieved through either one, or a combination, of the following activities:

- Ultrasonic testing (UT) wall thickness inspection of piping, pressure vessels, and tanks
- Corrosion rate monitoring through using weight loss coupons and electrochemical probes
- Intelligent pigging of transfer pipelines
- Deterioration rate measurements at failure and leak locations

Because MIC often manifests itself at the bottom (or the 6 o'clock position) of equipment, wall thickness inspections must include equipment bottoms. In addition, inspecting the lower half of all relevant equipment must be included in a site's overall inspection plans.

Similarly, any corrosion rate monitoring devices used or intended for MIC deterioration rate monitoring should be fitted at the 6 o'clock position. Alternatively, if they are fitted at the 12 o'clock position, their fittings should be long enough to enable the associated weight loss coupons or electrochemical probes to reach, or to be in close proximity of, the equipment bottoms.

It should be noted that the measured or estimated deterioration rates (through wall thickness inspections and corrosion rate monitoring devices) may be attributed to MIC if the assessment stage (or Stage 2) later revealed that bacteria or bacterial activity existed within the areas where inspections or corrosion rate monitoring had taken place and simultaneously, the required nutrition and bacterial growth conditions existed thereof. Otherwise, if no or little signs of bacteria or bacterial activity was discovered, then the recorded deterioration rates should be attributed to other possible (abiotic) deterioration mechanisms such as under deposit corrosion.

4.3.2.1 Equipment Wall Thickness Inspections

Carrying out wall thickness inspections provides the operator or asset owner with a means to measure or estimate the ongoing wall loss due to MIC, if the assessment stage later identified bacteria or bacterial activity at such locations. In other words, for locations where MIC (or more precisely biofilm or sessile) presence has been identified during the assessment stage, such inspections can help to determine or estimate the associated MIC deterioration rates.

Because MIC often manifests itself at the bottom of equipment, then such inspections must focus on these areas or at least adequately cover these areas as well. If the inspections being carried out are risk-based and the associated (corrosion) failure risk due to MIC has been accurately determined, then one of the outcomes of such failure risk assessment exercise will be the determination of locations that suffer from MIC. Thus, proper and thorough risk assessment campaigns can lead to the accurate determination of locations (mostly at the bottom of equipment) where MIC is taking place. Then, carrying out wall thickness inspections at such locations can help the operator determine the associated MIC deterioration rate and the remaining lives of the assets.

Similarly, if dead leg areas exist within the asset, then such locations are associated with a higher risk of MIC provided that the bacterial nutrition and growth conditions exist within these locations. Consequently, when creating inspection plans, dead leg areas need to be included in the pertinent inspection scopes.

4.3.2.2 Corrosion Rate Monitoring

Corrosion rate monitoring devices, such as weight loss coupons and electrochemical probes, can provide useful information about MIC deterioration rates for locations where the presence of sessile bacteria or biofilms have been reported. This is contingent on the rate monitoring devices having been fitted at the 6 o'clock position and being in direct contact with the existing aqueous phase.

The online electrochemical probes, if adequately and regularly maintained and serviced, can provide live or real-time information about the onset of MIC and the associated deterioration rates. Where the main or prevalent deterioration mechanism has been determined to be MIC, then any corrosion rates recorded or reported by such devices can be attributed to MIC.

4.3.2.3 Parameters Affecting MIC Rate Monitoring

Parameters that affect and influence the quality and accuracy of MIC rate monitoring include the following:

- Location of wall thickness inspections and corrosion rate monitoring devices
- Orientation or clock position of wall thickness inspections and corrosion rate monitoring devices
- Frequency of inspections and frequency of servicing the corrosion rate monitoring devices
- Whether or not inspections carried out are risk-based
- Identifying dead leg areas, or locations for inclusion in the upcoming wall thickness inspections and/or fitting of corrosion rate monitoring devices
- Competency of those personnel who carry out wall thickness inspections or manage and maintain the corrosion rate monitoring devices

4.4 Summary

The bacterial and MIC mitigation process is composed of the following three stages:

- Bacterial and MIC monitoring stage
- Bacterial and MIC assessment stage
- Bacterial and MIC control stage

As can be seen from above, the main three activities of the overall mitigation process (i.e., monitoring, assessment, and control) are all applied on two parameters: bacteria and MIC. In other words, and more precisely, the overall mitigation process is composed of three stages and six components.

The first stage of the mitigation process focuses on bacterial and MIC monitoring. Bacterial monitoring produces the necessary liquid and biofilm (or sessile or solid) samples for the bacterial assessment phase, which can be carried out either in the field itself or later in a lab. MIC monitoring can provide information about ongoing corrosion rates due to bacterial activity and the remaining lives for the equipment concerned.

To ensure that both components of stage 1 are carried out adequately and accurately, the various parameters that affect the application and performance of such components must be taken into consideration. These parameters include location, orientation, frequency, and competency of those who carry out such activities on a regular basis.

5 Bacterial and MIC Assessment

5.1 Introduction

Once the required (liquid and solid) samples are taken and wall thickness loss, weight loss, and corrosion rate data are collated, then they must be processed, trended, and analyzed. These are done to determine bacterial types, bacterial numbers, concentration of various chemical species, and the pertaining MIC corrosion rates.

The bacterial assessment component can be carried out using traditional (culture-based) methods or modern (culture-independent) ones. Lastly, one must determine whether or not the observed corrosion is due to the encountered bacteria. *This is where MIC diagnosis can help.*

5.2 Bacterial and MIC Assessment Stage

During this stage, the samples and data collected during Stage 1 are processed and analyzed. Like Stage 1, this stage also is composed of two components:

- Bacterial Assessment Component
- MIC Assessment Component

Throughout the bacterial assessment process, the liquid and solid samples taken during Stage 1 are analyzed to determine the following:

- Bacterial types
- Bacterial populations

- Presence and concentration of various chemical species (e.g., ions, molecules, and enzymes) produced and/or consumed by the existing bacteria
- Fluid composition
- Residual or concentration of various chemicals (e.g., chlorine or any of its derivative chemicals, oxygen scavenger, and biocide)
- Total dissolved solids
- Presence of any corrosion products and their composition

Simultaneously, wall thickness inspection data along with corrosion rate monitoring data are analyzed and trended to first determine or estimate the ongoing corrosion deterioration rate and then (using the results of the bacterial assessment component) to decide whether the calculated or estimated deterioration rates are due to bacterial activity or other abiotic deterioration mechanisms. (See Section 5.2.3 Diagnosing MIC, for more detail.)

5.2.1 Bacterial Assessment Component

Bacterial assessment refers to methods or techniques used to determine either the bacterial types, or the associated bacterial populations, or both. Such methods can be generally divided into the following two categories:

- Traditional or culture-based assessment methods
- New, modern, or culture-independent assessment methods

Sections 5.2.1.1 and 5.2.1.2 describe the above two categories, respectively, while Section 5.2.1.3 discusses some general facts about such assessment methods.

5.2.1.1 Traditional or Culture-Based Assessment Methods

Such methods are based on growing microorganisms in general–and bacteria in particular–on a solid or in liquid media and then counting them using color or turbidity changes, respectively. To culture bacteria, a small amount of liquid or a suspension of a solid is added to a solution or solid that contains nutrients (i.e., the culture media).

For solid samples, growth is detected as a color change (due to a chemical reaction). This indicates the formation of bacterial colonies on the surface of culture media after incubation for around 28 days.

Liquid media, however, are inoculated with a liquid sample taken from the field and can be incubated up to 14 days. The change in turbidity is then regarded as an indication of bacterial activity.

There are two main traditional or culture-based methods:

- Serial dilution method–used for liquid samples
- Most probable number (MPN) method–used for solid samples

The MPN method yields more accurate bacterial numbers compared to the serial dilution. To improve the accuracy of the estimated bacterial numbers for either method, more replicate samples can be used.

For the culture-based techniques, there are three main variables:

- Type of culture medium
- Incubation temperature
- Length of incubation

While the traditional or culture-based methods are, in general, cheaper than newer or modern techniques, as well as also easier to carry out, there are several shortcomings or limitations associated with such methods.

- Incubation at one single temperature is selective.
- Such methods are not applicable to stored corroded pieces.
- These methods are time consuming; while serial dilution can take up to 14 days to yield results, the MPN method requires 28 days.
- Culture media often do not approximate the natural environment.
- Some microorganisms or bacteria are not culturable, that is, they cannot be grown in culture media.
- Traditional methods might underestimate the true number of bacteria by at least one order of magnitude.
- A specific culture medium is only capable of growing the specific bacteria that can use or metabolize it as nutrient.
- The MPN method requires a large amount of space (for sample storage and incubation).

5.2.1.2 Modern or Culture-Independent Assessment Methods

Modern bacterial assessment methods are mostly culture-independent and do not require (liquid or solid) growth media for determining bacterial types and or bacterial enumeration. Because they are culture-independent, they are free from the inherent limitations and shortcomings that are generally associated with the culture-based bacterial assessment techniques (Section 5.2.1.1).

The modern bacterial assessment methods can be categorized into the following classes or types.

- Molecular microbiological methods (MMM): They are also called the "genetic-based" methods. Compared to the traditional culture-based methods, they provide a more precise enumeration of various bacterial types in a significantly shorter timeframe. Examples of such methods include
  - 4',6-diamino-2-phenylindole (DAPI)
  - Fluorescence In Situ Hybridization (FISH)
  - Quantitative Polymerase Chain Reaction (qPCR)
  - Next Generation Sequencing (NGS)

- Photometry-based method: This method is based on measuring the amount of the adenosine triphosphate molecule that is present in all living cells through photometry. The best example is the adenosine triphosphate (ATP) photometry method.
- Microscopy-based methods: Microscopy is used to count bacteria, which have been tagged with fluorescent dyes, stains, or other reagents that can be detected via various microscopic methods such as fluorescent microscopy.
- Enzyme-based methods: In this method, the total number or concentration of an enzyme (produced by a particular bacterium) is measured. The measured quantity then provides an indication of the concentration of a certain active bacterial type. The best example is the adenosine-5-phosphosulfate (APS) reductase assessment method, which can indicate the concentration of active SRB in the provided sample.

In general, the new culture-independent methods provide significantly faster results than the traditional culture-based methods; however, they are usually more expensive than the latter. Tables 5.1 and 5.2 in Section 5.2.1.3 list various attributes of the above methods (along with the culture-based methods for comparison purposes).

5.2.1.3 General Facts About Bacterial Assessment Methods

As was mentioned in Section 5.2.1 (The Bacterial Assessment Component), the bacterial assessment methods can be divided into two groups: culture-based (or traditional) and culture-independent (or modern) methods. Despite various advantages offered by using the modern methods, in many upstream oil and gas operations, the asset owners, operators, and contractors often still use the traditional ones. These are not only cheaper, but also easier to use and less training is required, than the modern culture-independent methods such as molecular-, microscopy-, and photometry-based assessment methods.

In addition to reduced cost and less training needed, another reason for the continued prevalence of the traditional methods may be that some of the principal attributes of the modern methods (see Table 5.1) are still somehow unknown to many asset owners and operators. Despite their higher costs and their requirements for sophisticated equipment and instrumentation, the modern bacterial assessment methods possess various advantages over the traditional ones, of which speed (i.e., time required to yield results) is one of the critical ones. Table 5.2 lists both traditional and modern bacterial assessment methods along with their pertinent advantages and disadvantages.

Following are some general facts about bacterial assessment methods.

- More than one assessment method is often required to fully and adequately appreciate the true effect of bacterial contamination associated with any upstream oil and gas asset.

- All methods have their own limitation, but have certain associated advantages and disadvantages.
- There is no relationship between the extent of MIC (i.e., corrosion and corrosion rate due to bacterial contamination) and the number or activity of bacteria identified via any of the aforementioned assessment methods.
- Sessile bacteria should definitely be included in any bacterial sampling and assessment activities (along with planktonic bacteria).
- MMM provide a powerful means of precisely monitoring the effectiveness of biocide treatments.[4]
- MMM were first broadly introduced into the oil and gas industry in 2007.[4]
- The qPCR remains one of the best monitoring tools available for the quantification of total SRB populations.
- Changes in temperature that occur during transit and storage can influence a change in the number and ratio of bacteria in the sample.[4]
- Samples for MMM should be stored and shipped on ice in less than 24–48 hours.[4]
- It should be remembered that the transport of samples from the field–prior to assessment–may significantly affect the type and population of the bacteria; therefore when practical, the assessment stage should be initiated in the field.[1]

5.2.2 The MIC Assessment Component

The raw data produced and collected during Stage 1 and its MIC monitoring component are converted into corrosion rate data, if they are not already in such format (i.e., millimeters per year [mm/y] or mils per year [mpy]). The best examples of such raw data are ultrasonic thickness (UT) wall loss data (in millimeters) and weight loss coupon data (in grams), which must be converted and processed into corrosion rate data (in mm/y or mpy).

Thus, during the MIC assessment component, all the pertaining corrosion rates are determined or estimated. Most of such corrosion rates are associated with the lower half of the equipment, or more precisely, the 6 o'clock position. However, one critical point to remember is that not all corrosion rates associated with the 6 o'clock position are necessarily due to MIC and they can be due to other abiotic corrosion deterioration mechanisms such as the under deposit corrosion (UDC) mechanism.

In general, corrosion rates due to MIC can be very severe and in the range of several millimeters per year. However, there is as yet no relationship between the bacterial population and the magnitude of corrosion. Therefore, one should not relate the quantified bacterial numbers (Section 5.2.1) to any measured or estimated corrosion rates. Where SRB have been identified, there is no relationship between the extent of MIC (i.e., corrosion) and the level or activity of SRB, or even bacteria in general.

Once the pertaining corrosion rates have been calculated or estimated, one has to determine whether such deterioration rates have been due to MIC or not. This is when MIC diagnosis (Section 5.2.3) plays an important role.

TABLE 5.1 General attributes of both traditional and new bacterial assessment methods.[1, 6]

Assessment Method	Base of Assessment Method	Where Assessment Can Take Place	Capable of Assessing				Output Type	Required Time to Produce Output
			Live Cells	Dead Cells	Liquid Samples	Solid (Biofilm) Samples		
Serial Dilution Method	Culture-based	Field & Lab	Yes	No	Yes	No	Bacterial number estimates	2 to 4 weeks
Most Probable Number (MPN)	Culture-based	Field & Lab	Yes	No	Yes	Yes	Bacterial number estimates	Up to 4 weeks
ATP	Culture-Independent & Photometry	Field & Lab	Yes	No	Yes	Yes	Bacterial number estimates	Up to 10 min
APS Reductase Measurement	Enzyme-based	Field & Lab	Yes	*Not indicated*	Yes	*Not indicated*	SRB number estimates	*Not indicated*
Fluorescent Microscopy	Microscopy	Lab	Yes	Yes	Yes	Yes	Bacterial number estimates	*Not indicated*
DAPI	Microscopy & MMM	Lab	Yes	Yes	Yes	Yes	Bacterial number estimates	Up to 12 h
FISH	Microscopy & MMM	Lab	Yes	No	*Not indicated*	*Not indicated*	Bacterial number estimates for the likes of SRB and SRA	*Not indicated*
NGS	MMM	Lab	Yes	*Not indicated*	Yes	Yes	Bacterial number estimates & bacterial types	Up to 1 week
Hydrogenase Measurement	Enzyme-based	Lab	Yes	*Not indicated*	*Not indicated*	Yes	SRB contamination	Between 4–12 h
qPCR	MMM	Lab	Yes	Yes	Yes	Yes	Bacterial number estimates, specific bacteria type number estimates	*Not indicated*

Not indicated: The information required could not be found or inferred from the two references used.

TABLE 5.2 Advantages and disadvantages associated with both traditional and new bacterial assessment methods.[1,6]

Assessment Method	Advantages	Disadvantages
Serial Dilution Method	Cheaper and easier than culture-independent methods	Refer to Section 5.2.1.1
Most Probable Number (MPN)	Cheaper and easier than culture-independent method; results more precise than the serial dilution method	Refer to Section 5.2.1.1
ATP	Very quick way of estimating bacterial numbers; no underestimation of unculturable bacteria; used to monitor bacterial trends (e.g., before and after the biocide treatment)	Should be backed up with at least one other bacterial quantification method; does not provide bacterial speciation; requires sensitive instrumentation
APS Reductase Measurement	Does not require bacterial growth in a medium; faster than culture-based methods	Only applicable to SRB
Fluorescent Microscopy	Only requires very little amount of sample; faster than culture-based methods	Interference possible with organic and inorganic materials present in the sample; lab-based technique and not applicable in the field
DAPI	Only requires very little amount of sample; faster than culture-based methods	Does not identify bacterial types or species
FISH	Specific bacterial types such as SRB can be targeted; detects archaea as well; faster than culture-based methods	Requires 50–100 mL of sample fluids; lab-based technique and not applicable in the field
NGS	High sensitivity to identify various bacterial types; faster than culture-based methods	Requires high-tech equipment
Hydrogenase Measurement	The presence of hydrogenase may indicate a potential for MIC	Requires high-tech equipment and delicate sample preparation; only applicable to SRB
qPCR	Capable of both quantifying the total number of or a specific bacterial type; can be applied to any type of sample (i.e., fluid and solid)	Lab-based technique and not applicable in the field

5.2.3 MIC Diagnosis

As discussed in Sections 5.2.1 and 5.2.2, the bacterial assessment and MIC assessment components facilitate the determination of bacterial types and numbers as well as the estimation or calculation of the associated corrosion rates, respectively. However, the mere presence of bacteria along with recording some corrosion rates does not necessarily indicate the occurrence of MIC for sure. To both correctly and accurately diagnose that MIC has been taking place, all the following items are required:[5]

- A corrosion product sample
- Identification of a corrosion deterioration mechanism pertaining to the metal being examined
- Identification of bacteria or other microorganism forms capable of growth, survival, and maintenance of corrosion deterioration mechanism in the particular environment being studied
- Demonstrating that 1, 2, and 3 are related or associated with each other

In other words, the MIC diagnosis process entails the interpretation of the following simultaneously:[5]

- Chemical data (e.g., presence of certain chemical species, pH, and an increase or decrease in the concentration of various specific molecules, cations or anions, and suitable growth conditions)
- Metallurgical data (e.g., corrosion rate, presence of corrosion product and the pertaining corrosion deterioration mechanism, deterioration or failure location morphology, and failure orientation [i.e., the clock position of the failure])
- Microbiological data (e.g., bacterial types, presence of biofilm, and bacterial numbers)

Last but not least, the pit morphology by itself, should not be used for MIC diagnosis alone, because such features could also be produced by abiotic reactions.[5]

5.3 Summary

The second stage of the bacterial and MIC assessment stage uses the raw data produced during Stage 1 to identify bacterial types and various chemical species, to estimate the associated bacterial numbers, and finally to estimate or calculate the encountered corrosion rates. Diagnosing actual MIC would help to determine whether the encountered corrosion and the measured or estimated corrosion rate are due to the observed bacterial activity or not.

During the bacterial assessment component, one has the option to use either traditional bacterial assessment methods or modern ones, or both.

6 Bacterial and MIC Control

6.1 Introduction

Bacterial and MIC control is the third and the last stage in the overall bacterial and MIC mitigation process. Unlike the previous two stages of this process (Chapters 4 and 5) it is rather difficult to have two distinctive components here; that is, one for bacteria and another for MIC-related activities. This is because through implementing bacterial control activities one can simultaneously control MIC and vice versa.

Optimally, bacterial and MIC control should begin during an asset's design phase. Appropriate and correct design, and materials selection choices and considerations throughout this phase will have a profound influence on the effectiveness of the overall bacterial and MIC mitigation process postcommissioning.

A very good example is the elimination–as much as possible–of dead leg areas during an asset's design phase. Eliminating these would have significant effects on the bacterial and MIC control activities later and during the asset operation phase.

The bacterial and MIC control stage is composed of various activities that range from water removal (or product dehydration) to chlorination, biocide treatment, pipeline cleaning pigging, and nitrate treatment. The choice of the most appropriate activity or treatment will always depend on a number of factors:

- Available resources (time, money, and personnel)
- The nature, extent, and severity of the existing or predicted bacterial contamination and MIC issues

- The available hardware (e.g., sampling and chemical injection points)
- Competency of the relevant asset managers and personnel

Based on the author's personal experience and observations, proper application of corrosion management principles also has a significant influence on the effectiveness of any bacterial and MIC mitigation process. Therefore, when creating, revising, or updating a bacterial and MIC mitigation methodology or strategy for any particular asset, corrosion management applications should definitely be included in any such methodology or strategy.

6.2 Brief Introduction to the Bacterial and MIC Control Stage

The bacterial and MIC control stage is the third and final phase or stage in the overall bacterial and MIC mitigation process. Each of the two previous stages (i.e., monitoring and assessment) were divided into two separate and rather discrete components–the first one always focusing on bacterial-related activities or measures, and the second one on MIC-related ones. However, the activities or measures pertaining to the bacterial and MIC control stage cannot be similarly divided into two separate components or categories. This is mainly because the boundary between the bacterial control activities and the MIC control activities is rather blurred and indistinguishable because by controlling (or mitigating) bacterial activity, growth, and metabolism an operator is also simultaneously controlling (and mitigating) MIC, and vice versa. Consequently, this final stage will be presented under the single category of "bacterial and MIC control activities" hereinafter.[1]

The technical expertise and guidelines presented throughout this chapter are of general nature and common to most operation types where MIC is an integrity issue. Therefore, such expertise can be adapted by most process and utility system types susceptible to MIC. However, the contents of this chapter might be more suitable, relevant, and useful for the types of assets and operations that have an inlet water system through which they receive raw (i.e., untreated) water from sources such as

- Oceans and seas
- Rivers
- Wells and aquifers

Inlet water systems usually rely on a two-stage bacterial and MIC control system–chlorination followed by biocide treatment. More precisely, such inlet water systems use an oxidizing chemical (such as chlorine) upstream of the deaeration unit followed by injecting a nonoxidizing chemical (such as a biocide) downstream of the deaeration unit.

1. Unlike the previous two stages (described in Chapters 4 and 5) the pertinent activities will not be categorized into two discrete components (i.e., the bacterial control activities and the MIC control activities).

While most control measures covered in this chapter pertain to the asset operation phase, the significance of carrying out proper and adequate design and materials selection considerations and reviews (from a bacterial and MIC viewpoint) during the asset design phase should not be ignored or underestimated. Such an approach can significantly improve the efficiency of any bacterial and MIC control measures implemented during the asset operation phase. Relevant design and materials selection advice and guidelines are presented in Sections 6.8.2 and 6.8.3.

The following sections cover various bacterial and MIC control measures and activities, which have been used across various assets and operations around the world.

6.3. Chlorination

Antimicrobial chemicals are broadly divided into two groups–oxidizing and nonoxidizing ones. The oxidizing group includes chemicals such as chlorine (Cl_2), bromine (Br_2), and sodium hypochlorite ($NaOCl$). Such oxidizing chemicals are inorganic compounds, which possess very high chemical activity and are thus short-lived. They are the best option in treating very large amounts of water at a relatively low cost. Compared to nonoxidizing chemicals, they have a low environmental impact.

The chlorination treatment can be defined as a chemical treatment based on the *continuous* injection of molecular chlorine (an oxidizing chemical) or similar oxidizing compounds such as sodium hypochlorite into the inlet of water systems.

Upon injection into water, molecular chlorine reacts with water and produces hypochlorous acid ($HOCl$) and hydrochloric acid (HCl). The former ($HOCl$) is a weak acid and further dissociates to form hydrogen and hypochlorite ions per the following reactions:

$$Cl_2 + H_2O \rightarrow HCl + HOCl \qquad \text{(R.6.1)}$$

$$HOCl \leftrightarrow H^+ + OCl^- \qquad \text{(R.6.2)}$$

The main oxidizing agents in water are $HOCl$ and OCl^- (see Reaction 6.2 above). Under certain conditions (i.e., the right temperature and acidity), $HOCl$ is nearly two orders of magnitude more effective than the hypochlorite ion, in killing (planktonic) bacteria. A free residual chlorine of up to 0.5 ppm is deemed adequate to control and inhibit planktonic bacteria if such residual concentration can be maintained for a sufficient amount of time. For effective chlorination, a residence time of up to 30 min might be required to achieve complete (planktonic) bacteria kill. Increasing the temperature and the residence time will increase the efficiency of the kill or (planktonic) bacterial inhibition.

Below are some of the most important points to be considered throughout chlorination campaigns.

- Chlorination is not a batch or intermittent chemical treatment–chlorine or any similar oxidizing chemicals should be applied *continuously* at a low dose.
- If electrochlorinators (used to produce hypochlorite out of seawater) are used for chlorination purposes, they require regular maintenance to ensure both smooth and continuous chlorination at the water inlet.
- The residual concentration of chlorine should be monitored regularly. Such monitoring can be done using either online methods or portable kits.
- The strong oxidizing nature of chlorine means that it will also react with any bare carbon steel substrate as well as the oxygen scavenger chemical (if there is any such chemical in the system). Therefore, chlorination should only be practiced upstream of the deaeration unit (if there is one) where ideally no bare carbon steel and oxygen scavenger chemical should be present.
- Chlorination, when properly carried out, can only kill *planktonic* bacteria. It is almost totally ineffective against sessile bacteria and biofilms because chlorine is inactivated by the organic matter surrounding the sessile bacteria.
- The effectiveness of the chlorination treatment is inversely proportional to pH. For example, at pH: 8.0– 9.0 the chlorine may only be 1–10% active.
- Hypochlorite solutions are not completely stable during storage time and can lose 25% of their activity or strength after several months in storage.
- The presence of reducing agents in the water phase will reduce the efficiency of the chlorination campaign.

6.4 Biocide Treatment

6.4.1 Liquid Biocide Treatments

Liquid biocides are nonoxidizing chemicals that exert their effect by reaction with specific cell components or reaction pathways in the cell. They are often added into the system as batch treatments and (most often) downstream of the deaeration tower in seawater treatment systems.

Common classes of biociding chemicals include

- Aldehydes
- Phenols
- Imidazolines
- Amines
- Tetrakis (hydroxymethyl) phosphonium sulfate (THPS)

The most common biocide type is the solution of glutaraldehyde blended with surface active agents. Surface active agents are needed for penetration into biofilms or sessile colonies already formed on the surface. Quaternary amines are typical surface active agents added into the nonoxidizing biocide chemicals.

Unfortunately, a common mistake or shortcoming is the purchase of nonoxidizing biocide chemicals that lack surface active agents. In such cases, the purchased chemical may only kill planktonic bacteria and it is totally ineffective against sessile bacteria often existing within biofilms. The main reason they are ineffective against sessile bacteria is that in the absence of such surface active agents, the associated chemicals are incapable of penetrating into the biofilms to kill or destroy the sessile bacteria entrenched there. Thus, the biofilm structure offers protection against such chemicals that lack surface active agents.

Therefore, it is very important to ensure that the chemical or biocide to be purchased already contains surface active agents and is capable of biofilm penetration and killing sessile bacteria. Even if the chemical supplier produces a lab certificate stating that a chemical is capable of killing 100% of the bacteria, as demonstrated in the conducted lab tests, the certificate has to clearly and transparently specify that the tested chemical is capable of "*killing 100% of the sessile bacteria*" and not just "*bacteria.*" The latter often refers to *planktonic* bacteria only; however, this fact sometimes is not appreciated by potential buyers. Therefore, the general lab certificate statement that "this chemical kills 100% of the bacteria" is erroneously interpreted and understood by many buyers as if the chemical is capable of killing both planktonic and sessile bacteria.

Below are some general facts about nonoxidizing biocides.

- Quaternary ammonium salts are only used downstream of the gas stripping towers (associated with seawater injection systems) because of their high foaming tendency.
- Variation in pH has little effect on the chemical nature of glutaraldehyde; however, the effect of pH on bactericidal action is great. It is noticeably more effective under alkaline conditions than under acidic conditions.
- Although it would be wise to add biocide directly to the inlet of the stripping tower to maintain an efficient and clean deaeration tower (often associated with seawater injection systems), practical considerations (such as foaming) may prevent this option.
- Traditional dose rates are approximately 200 ppm for six hours every week (depending on MIC severity).
- The new practice is a higher dose rate of 1,000 ppm or more for one hour at an appropriate time interval. This procedure uses less chemical and is also more effective. Monitoring should enable the operator to identify the optimum treatment.
- After biocide treatment, it is not unusual to see a rise in SRB for a short time as the biocide removes bacteria from the surface, i.e., planktonic bacterial numbers may rise after each biocide treatment.
- The biofilm layer is not easily penetrated by nonoxidizing biocides. Therefore, surface active agents should be blended in the formulation of such chemicals or biocides to facilitate biofilm penetration.
- Oxygen scavengers must be added upstream of biocides (because of potential incompatibility) and should not be overdosed (because of corrosion risk).

- In general, cleanup of a fouled system requires a higher concentration of intermittently fed treatment, while the maintenance of a clean system can be achieved with low-level continuous or semicontinuous feed of biocides.
- Biocide injection into the deaeration tower (associated with seawater injection systems) compromises deaeration during biociding.

The main attributes of the nonoxidizing biocides include the following:

- Short organic molecules
- Less reactive than oxidizing ones
- Lower reactivity allows improved penetration; thus, enhanced killing of *sessile* bacteria, which are entrenched within biofilms
- Function by disrupting the cell metabolism
- Aldehyde-based ones have severe health, safety, and environmental issues
- Severe handling risks

Nonoxidizing biocides *should ideally possess* the following properties:

- Be highly effective against a wide range of bacteria at low concentrations
- Be noncorrosive (at least against carbon steel) and safe to handle
- Be as specific as possible for the existing bacterial types and should not become inactivated by other materials (i.e., either by other injected chemicals or by the polymeric material associated with biofilms or both)
- Have good penetration properties, enabling them to thoroughly penetrate through the biofilm and reach the (sessile) bacteria at the metal surface
- Be stable during storage and should not become inactive or less efficient
- Be readily available at low cost
- Be completely bactericidal (i.e., killing bacteria) within a short time and not inducing bacteriostasis (i.e., inactivating bacteria instead of killing them)
- Not be toxic to animals and plants

Biocide testing, selection, and purchasing is a critical and elaborate process. Even some trivial mistakes, oversights, or shortcomings can lead to purchasing the wrong chemical with no or little bactericidal properties.

Following are some of the observed and common mistakes during biocide effectiveness testing:

- Varying quality and composition of the cultures used
- Varying incubation temperature and times
- Testing against pure bacterial types
- Testing against planktonic bacteria only. This is, unfortunately, a common issue and some operators purchase biocide chemicals which were only tested against planktonic bacteria

Unfortunately, some unscrupulous individuals or chemical suppliers market and sell so-called biocide chemicals using tricks or underhanded methods. These often mislead the operators through convincing them to purchase inefficient and non-bactericidal chemicals. Some of such common tricks include the following:

- Marketing and selling *planktonic-specific* biocide chemicals as those capable of killing *sessile* bacteria as well
- Marketing and selling biocide chemicals with a variable active component; thus, varying the killing power
- Marketing and selling biocide chemicals that lack surface active agents; thus, they are incapable of adequate penetration into biofilms to kill sessile bacteria

In general, factors affecting biocides' killing power and rate of killing include

- Biocide concentration (i.e., the higher the applied concentration of the biocide chemical the higher the killing rate)
- System temperature (where the biocide is injected and enters the system)
- Two-phase systems (i.e., how the biocide chemical is partitioned between the two existing phases)
- Presence of organic matter (i.e., the polymeric matter associated with biofilms that also acts as a barrier for any applied biocide chemical)
- Chemical deactivation by other injected chemicals
- Population size of the existing bacterial colonies
- Environment pH
- Duration of application or chemical injection into the system

6.4.2 Liquid Biocide Activity

Biocide activity is often expressed in percentage and can be defined as the amount of active substance(s) or chemical(s) in grams per 100 mL of (biocide) solution. For example, a biocide with 40% activity has 40 g of biocidal (or inhibiting or killing) substance(s) dissolved in 100 mL of solution. The higher the activity of a biocide, the greater is its inhibiting or killing potential. Simultaneously, biocides with higher activity are likely to be more expensive than similar biocides with lower activity values.

During lab testing, field trials, and field applications the biocide activity values should be clearly specified in all pertaining documentation, correspondence, and procedures. Otherwise, any ambiguities associated with the exact value of biocide activity or failure to mention or specify it, can lead to confusion over the exact or relevant biocide efficiency. Such ambiguities can also cause potential repercussions concerning proper and adequate bacterial and MIC control and mitigation leading to more severe and frequent failures due to MIC.

The associated issues pertaining to unspecified biocide activity values can be best demonstrated with the following example:

Two different chemical suppliers produce, market, and sell a virtually identical biocide, but under two different brand names. For the sake of simplicity, the biocide produced by the first supplier is called *Biocide A* and the one produced by the second supplier *Biocide B*. The activity values of biocides A and B are 20% and 50%, respectively. However, the activity values are not specified at all, so any potential buyer would be completely unaware of the associated activity values.

Biocide A is significantly cheaper than Biocide B, because considerably less active substance(s) (20% vs. 50%) has been used in its production; this is assuming that both biocides use more or less the same type of solvent or carrier liquid(s) with almost identical costs.

The first supplier is recommending its Biocide A to be injected every 10 days at 400 ppm, while the second supplier is recommending 200 ppm at the same time interval. Considering this scenario and the available options, most asset owners and operators would prefer Biocide A over Biocide B for the following reasons:

- At the recommended concentration of 400 ppm, Biocide A appears to be more effective than its rival with the recommended concentration of 200 ppm.
- More importantly, Biocide A is conspicuously cheaper than the other product (because of the presence of less active substance(s) in it) and cost is a very important factor for all asset operators when considering the purchase of a new chemical.

However, in reality, the recommended bacterial treatment with Biocide B is going to be 20% more effective compared to the recommended treatment with Biocide A as shown below:

Biocide A actual activity: 400 ppm × 20% = 80 ppm

Biocide B actual activity: 200 ppm × 50% = 100 ppm

In summary, when dealing with biocides, their (chemical) activity should be clearly specified to further facilitate better decision-making regarding any required biocide treatment strategies or procedures.

6.4.3 Solid Biocide Treatments

In some cases, the threat or likelihood of bacterial activity and MIC is deemed low. This can be due to several reasons, but one of the most common ones pertains to new assets or equipment that have just been commissioned and no bacterial activity or MIC issues have yet been detected therein. However, it is inevitable that the threat of bacterial contamination and MIC will increase with time and they will manifest themselves at a foreseeable future. In such cases, to further delay or retard the advent of MIC, if the conditions permit, one can use solid biocide pills or rods, which are inserted, placed, or dropped into the associated equipment or systems that contain water. The solid biocide possesses very low dissolution rates (in water) thus providing the pertaining equipment and systems with a constant low dose of biocide over a certain period of time, which can last for a few years (depending on the amount of solid biocide and its dissolution rate).

The best example is a newly commissioned offshore diesel tank that regularly receives diesel from supply ships. The delivered diesel is often contaminated with microorganisms, and also contains some water. Such conditions can lead to increasing microorganism activity and MIC at the diesel–water interface at the bottom of the tank over time. However, by inserting or dropping a solid piece of biocide into the tank bottom, the MIC threat can be maintained at a lower or negligible levels thanks to the slow dissolution rate of solid biocide in the water phase thereby inhibiting bacterial activity, growth, and metabolism as long as any solid biocide exists at the bottom of the tank.

6.5 Water Removal (or Product Dehydration)

As mentioned in Section 3.3.1 (Bacterial Nutrition and Growth Conditions) water comprises more than 80% of the bacterial cell weight. Thus, water can and should be regarded simultaneously as both the main bacterial constituent and nutrition.

In addition, an aqueous phase is always required to facilitate the transfer of bacterial nutrition and excrement to and from bacteria, respectively. Therefore, total absence of water is equal to no bacterial activity or zero MIC risk. Consequently, the more efficiently water removal or product dehydration is carried out, the more likely that bacterial activity, growth, and metabolism are inhibited, leading to enhanced bacterial control and MIC mitigation.

In the upstream oil and gas sector, water removal (or product dehydration) is carried out through either, or a combination, of the following processes:

- Oil or condensate water separation–often done through using slug catchers, separator vessels, hydrocyclones, and flotation units.
- Gas dehydration–often conducted via slug catchers, knockout drums, and injection of water-adsorbing chemicals such as methanol and glycols.
- Product purity and cleanness control–done through stringently adhering to various production, storage, and transfer protocols for the pertaining hydrocarbon products being produced, such as diesel.
- Pipeline cleaning pigging–carried out through regular or ad hoc cleaning pigging of the relevant pipelines to remove both the dropped-out water and any existing biofilms or sediments under which bacteria may prosper, grow, and metabolize (for more details see Section 6.6).

In brief, minimizing water content or water cut of a liquid hydrocarbon phase, or decreasing the water dew point of a gas phase, reduces bacterial and MIC risk in general. That is, the amount of water accompanying the hydrocarbon phase is directly proportional to the level of bacterial and MIC threat.

6.6 Pipeline Cleaning Pigging

Pipelines that carry water as the main product (such as seawater transfer pipelines) or by-product (such as wet oil or wet gas transfer pipelines) are susceptible to bacterial contamination, biofilm growth, and ultimately MIC, provided that

- Suitable (bacterial) growth conditions (such as stagnant zones, low points, or slow flow rates) exist
- Adequate nutrition is also present within the pipeline or some parts, or sections of it thereof

The pipeline cleaning pigging can then be considered as a physical way of

- Either sweeping or removing the (often dropped-out) water phase, or
- Scraping the biofilm or any associated corrosion product (such as iron sulfide scales), or
- Removing both the water and biofilm simultaneously

Depending on the extent and severity of biofilm activity, growth, and presence along the associated pipeline, some operators may opt to combine pipeline pigging with biocide treatment; that is, combining a physical (bacterial) cleaning method with a chemical one. To do so, the operator may use two consecutive cleaning pigs in which the space (or more precisely the volume) between them is filled with a slug of nonoxidizing chemical biocide.

In such a combined operation, the first pig physically removes any existing water, debris, biofilm, sediment, and corrosion product and thus provides better access for the biocide toward the bacteria that are entrenched within the associated corrosion sites. The second pig then further smears the biocide onto the pipeline wall and its various cavities, grooves, and pits, which might have been created by MIC or any other (abiotic) corrosion deterioration mechanism(s); located mostly at the bottom of the pipeline.

For any pipeline (that carries water either as the main product or as the by-product) there has to be a written cleaning pigging procedure clearly specifying the type of pig used, the pigging frequency, and any other relevant details such as the requirement to collect, assess, and analyze any pigging debris at the end of each pig run.

6.7 Nitrate Treatment

Nitrate treatment is a type of chemical treatment based on the *continuous* injection of a nitrate compound (such as calcium nitrate) into a water injection system. The main aim is to promote the activity, growth, and metabolism of rather benign bacteria such as the nitrate utilizing bacteria (NUB) or the nitrate-reducing bacteria (NRB) at the expense of SRB. By doing so, NUB and/or NRB consume the existing or the available nutrition also required for the growth of insidious bacterial types such as SRB. NUB and NRB grow much faster than SRB. Therefore, SRB activity and growth is restricted and mitigated both across an asset and within the associated reservoir(s). Simultaneously, the nitrate anion itself reacts with H_2S and removes it from the system. That is, nitrate can be regarded as a chemical scavenger for hydrogen sulfide.

The suppression of SRB activity and growth therefore leads to reduced risk of MIC across the asset and it also leads to a decreased likelihood of reservoir souring in the pertaining reservoir(s). However, nitrate treatment has so far *not* evolved into a worldwide and common practice or chemical treatment against the threat of MIC (due to SRB), and also against the occurrence of reservoir souring. Nitrate treatment has received some mixed reviews so far. Some of the points raised include the following.

- NUB convert nitrate to nitrite and the latter promotes pitting type corrosion in the associated pipelines.
- Some NRB types also promote MIC themselves.
- The success and efficiency of nitrate treatment with regard to inhibiting or mitigating reservoir souring depends on parameters such as reservoir characteristics, water breakthrough times, and nutrient availability.[7]

6.8 Other Bacterial and MIC Control Measures

This section covers various other measures which when implemented or adhered to can further improve bacterial and MIC control. Of prime importance are those measures such as design (e.g., shape of each piece and the existence of drain holes within certain equipment) and materials selection requirements (Sections 6.8.2 and 6.8.3), which should be taken into consideration during the design phase of any asset and its various equipment. Such measures play a crucial role in bacterial and MIC control postcommissioning.

Other equally important measures are those that pertain to *corrosion management applications* and simultaneously play a significant part in efficient bacterial and MIC control activities. Such measures are covered in Section 6.8.5.

6.8.1 Maintaining Continuous Operations or Avoiding Intermittent Production

As previously discussed in Section 3.3.1.2, low flow or stagnant conditions further facilitate and promote bacterial settlement on the bottom of equipment and the possible eventual formation and growth of biofilms leading to MIC.

A parameter or variable that promotes the above favorable condition is intermittent production cycles or flow regimes. During such cycles and when the operation has stopped (inducing near or zero production or flow rates), planktonic bacteria are more likely to settle to the bottom of internal surfaces to initiate biofilm growth leading to MIC. To avoid such likelihood, it is therefore strongly advised to maintain continuous operations and to avoid intermittent production–unless the systems, equipment, or parts concerned can be fully drained. Complete and proper drainage of equipment significantly reduces the probability of biofilm growth and the initiation of MIC because such activity (i.e., proper drainage) deprives bacteria of the required nutrition and favorable growth conditions.

6.8.2 Improved Design

The application of appropriate and correct design considerations during the design phase of an asset and its equipment can significantly reduce the likelihood of bacterial activity, growth, and metabolism as well as the associated MIC threat during the asset's operations phase or postcommissioning.

Some of the most important design principles that have to be taken into consideration during the asset design phase include the following:

- Avoid overdesign (from a size point of view) options leading to very low flowrates
- Design proper drainage facilities for equipment
- Design out low flow and stagnant areas
- Design for improved drainage in tanks
- Maintain coatings on the interior of storage tanks
- Design and include appropriate injection facilities

6.8.3 Improved Materials Selection

Similar to design considerations, proper and reasonable materials selection choices during an asset's design phase can significantly reduce the likelihood of bacterial growth and the associated MIC, postcommissioning.

Below are some of the basic facts pertaining to materials selection decisions and their influence on bacterial activity and MIC during an asset's operation phase.

- Using a material's pitting resistance equivalent number (PREN) to estimate relative resistance to MIC. The higher the PREN value associated with a particular metal or alloy, the higher is its resistance toward MIC. In other words, the susceptibility of a metal or an alloy is inversely related to its PREN value.
- MIC resistance is related to a material's ability to resist localized corrosion and specific biotic corrosive species such as sulfide, and organic acids such as acetic acid.

- MIC is frequently reported for carbon steel alloys and less frequently for stainless steel and duplex alloys.
- Copper-nickel (Cu-Ni) and Ni alloys have been successfully used in flowing aerated seawater, but MIC has been reported in some cases, particularly where flow was stopped for extended periods of time.
- Corrosion resistant alloys (CRA) or carbon steel-cladded CRA are selectively applied where MIC threat has been deemed high.
- Systems such as produced water, seawater, and fire water are highly susceptible to bacterial activity and MIC. Use nonmetallics where pressure, stresses, and fire-resistance requirements permit alternatives to metals.
- Apply internal coating for dead legs or areas and systems that are infrequently operated.

6.8.4 Proper Pipeline Hydrotesting

If pipeline hydrotesting is carried out according to a correct and proper procedure, then the MIC threat during and after the hydrotesting campaign is minimized. However, many hydrotesting campaigns are not carried out properly and such practices significantly increase the likelihood of bacterial activity and the associated MIC both during and after hydrotesting. Some of the main actions that minimize the MIC threat during and after hydrotesting include the following:

- Adding biocide to the hydrotest water
- Completely removing the hydrotest water (the dewatering phase) at the end of the (hydrotest) operation mostly via several pig runs (if possible)
- Properly and adequately filtering out macrofouling from the hydrotest water (which can turn into future bacterial nutrition upon decomposition; in particular, during prolonged hydrotest campaigns)
- Completely drying the pipeline after the dewatering phase using either dried air, dry nitrogen, and/or glycol or methanol plugs

6.8.5 Enhanced Corrosion Management Applications

The *asset corrosion management* or simply the *corrosion management* discipline is a subset of the *asset integrity management* topic whose focus is to deal with and mitigate the corrosion threats associated with an asset from the asset's design phase to fabrication, storage, transportation, installation, operation, and decommissioning phases. Other components or subsets of an asset integrity management system (AIMS) deal with other pertinent integrity threats such as natural disasters, third party damages, and erroneous operations (Section 2.2).

Corrosion management is a rather new concept, and its advent goes back to the UK's offshore oil and gas industry in the early 2000s. Because of its freshness and newness, the corrosion management concept is still (unfortunately) subject to various definitions and even different applications or field implementation methodologies, procedures, and instructions–with sometimes contradictory outcomes.

In this book, however, the preferred or adopted corrosion management concept definition is based on the Morshed Corrosion Management Model (MCMM).[2] According to the MCMM, the corrosion management concept is defined as

> "The process of carrying out integrity reviews for the required integrity management measures, regularly monitoring their performance, and assessing their effectiveness post-commissioning."[2]

Another equally critical definition associated with the MCMM is the *integrity management measure* concept. According to the MCMM, an integrity management measure is defined as

> "Anything (e.g., an activity, document, procedure, and an individual's competence) that contributes to an AIMS and influences its functioning, effectiveness, and outcome."[2]

Based on this definition, while design and materials selection are considered as integrity management measures, so are parameters such as personnel competency, communications, corrosion control matrices, and risk-based inspection.

A very crucial point to remember is that *corrosion management* and *corrosion engineering* are not the same concept or application. More precisely, corrosion engineering is a subset of corrosion management.

A corrosion management approach compared to a corrosion engineering one, not only provides significantly more options to identify the existing (or even foreseeable) asset integrity threats (from a corrosion viewpoint), but it also offers a greater range of solutions to improve and enhance the functioning and efficiency of an existing AIMS.

Some of the components or constituents of MCMM which can significantly improve bacterial and MIC control are covered in Sections 6.8.5.1–6.8.5.8.

6.8.5.1 Corrosion Control Matrices (or Plant Integrity Windows)

For almost any oil and gas asset type there are various physical and chemical variables that if kept within a predefined range, below, or above an acceptable threshold, the corrosion or the deterioration rate is minimized. Examples of such physical and chemical variables include

- Physical variables: temperature, pressure, flow rate, potential (i.e., voltage), water cut, and dew point

- Chemical variables: pH, biocide (or any other injected chemical) concentration, chloride residual concentration, and bacterial populations (pre- and post-biociding) per unit surface or volume or both

For each variable, at least the following five parameters must be determined:

- Activity
- Responsible person
- Location (where the activity takes place)
- Frequency (of the activity)
- Predefined range or acceptable threshold

Once all the above parameters are determined or defined for the associated physical and chemical variables, then grouping them together will produce a matrix (or table) containing exact information and instructions that are absolutely crucial for maintaining the integrity for a particular system. Putting together all the matrices for all the associated systems creates a set of matrices, which are referred to as the "corrosion control matrices" (or plant integrity windows). Such matrices impart further clarity and transparency into any existing AIMS and they are indispensable.

Ideally, corrosion control matrices should be compiled or created at the asset's design phase after all the pertinent operational, process, and chemical injection parameters have been determined and finalized. Once the asset has been commissioned and is in its operational phase, then the asset corrosion engineer should carry out the first integrity review process to fully review the matrices and revise them if necessary.

Corrosion control matrices play a significant and critical role in maintaining asset integrity in general and in efficient control of bacterial activity and MIC in particular–provided that they are compiled and implemented properly. Table 6.1 is an example of a corrosion control matrix compiled for a typical seawater injection system. The listed numerical ranges and values in Figure 6.1 might not apply to other projects and should not be adopted without prior reviews and appropriate lab tests.

TABLE 6.1 A common corrosion control matrix for a typical seawater injection system (u/s = upstream; d/s = downstream).

Activity No.	Responsible Person	Activity	Location	Frequency	Threshold
1.1	Offshore Chemist	Monitor residual chlorine levels	u/s of deaeration (DA) Tower	Daily	0.5–1.0 ppm
1.2	Offshore Chemist	Monitor residual chlorine levels[a]	d/s of DA Tower	Daily	0.2–0.3 ppm
1.3	Offshore Chemist	Monitor antifoam injection rate	u/s of DA Tower	Daily	Threshold not practical to measure, offshore chemist to monitor tank levels to ensure correct dosing
1.4	Offshore Chemist	Monitor biocide injection rate	u/s or d/s of DA Tower	Weekly	180 ppm
1.5	Offshore Chemist	Monitor scale inhibitor injection rate	u/s of DA Tower	Daily	No comment
1.6	Offshore Chemist	Monitor oxygen scavenger injection rate	d/s of DA Tower	Daily	10 ppb
1.7	Offshore Chemist	Monitor residual oxygen concentration (online)	d/s of DA Tower	Every six hours	10 ppb
1.8	Offshore Chemist	Monitor residual oxygen concentration (CHEMets)	d/s of DA Tower	Every six hours	10 ppb
1.9	Offshore Chemist	Monitor residual bisulfite concentration	d/s of scavenger injection point	Daily	0.64–1.28 ppm
1.10	Offshore Chemist	Monitor pH[b]	d/s of DA Tower	Weekly	7.0
1.11	Offshore Chemist	Check oxygen probe membrane	d/s of DA Tower	Daily	N/A
1.12	Offshore Chemist	Monitor residual oxygen concentration	d/s of Injection Pumps	Daily	10 ppb
1.13	Offshore Chemist	Calibrate online oxygen probe	d/s of DA Tower	Weekly	N/A
1.14	Offshore Chemist	Monitor sessile bacteria population[c]	d/s of Injection Pumps	Weekly: Pre- and post-biociding	10^2 per cm^2
1.15	Offshore Chemist	Monitor planktonic bacteria population	d/s of DA Tower	Weekly	1 per cm^3

a. Only where the deaeration is carried out through gas stripping

b. Overdosing of oxygen scavenger and/or use of stripping gas containing CO_2 and H_2S gases can render the treated water pH acidic

c. Preferably using a sidestream located d/s of the DA Tower

6.8.5.2 Corrosion Key Performance Indicators

Corrosion key performance indicators (corrosion KPIs) are key corrosion control matrices activities that can be used as indices or indicators to monitor the performance and effectiveness of an AIMS. Regular monitoring of their performance will immediately reveal whether an AIMS has been functioning satisfactorily or not. Corrosion KPIs are also regarded as an indispensable tool in maintaining and improving asset corrosion management.

Normally, any corrosion control matrices document is composed of many different activities which, when carried out on a regular basis, contributes to the overall AIMS. However, some of these activities play a more critical role in mitigating corrosion and maintaining the integrity of the asset. The performance of such activities can, in general, better reflect the successes and the shortcomings of the AIMS.

The relevant corrosion KPIs should always be selected from the activities listed in a corrosion control matrices document. Selecting the most appropriate corrosion control matrices activities as corrosion KPIs is an exacting job and is normally at the discretion of the responsible asset corrosion engineer. Once the appropriate corrosion control matrices activities have been selected as corrosion KPIs, they will usually be calculated daily (provided that the associated activities are carried out on a daily basis) and expressed and reported as "percentage compliance on a monthly basis," preferably within the asset monthly corrosion report.

Normally, a compliance target level is also defined and used as a benchmark to evaluate the overall performance of the calculated monthly average compliance. A spreadsheet can be used to perform all the necessary calculations to determine individual KPI levels, their monthly average, and to plot the required graphs.

Table 6.2 depicts a typical KPI table taken from an asset monthly corrosion report where each individual activity is listed along with its corresponding range or threshold, monthly performance, and the compliance target level.

TABLE 6.2 A typical corrosion KPI table pertaining to a seawater processing facility.

PERFORMANCE MEASURE	FREQUENCY	TARGET VALUE/RANGE	% COMPLIANCE
Monitor residual chlorine levels	Daily	0.5-1.0 ppm	85
Monitor biocide injection rate	Weekly	180 ppm	90
Monitor oxygen scavenger injection rate	Daily	10 ppb	100
Monitor residual bisulfite concentration	Daily	0.65-1.30 ppm	65
Average Compliance			85
Compliance Target Level			80.0

Corrosion KPIs pertaining to bacterial activity, MIC and the associated mitigation activities such as water removal, regular pipeline cleaning pigging, and biocide treatments can play a significant role in both improving the awareness regarding the efficiency of the existing bacterial and MIC mitigation system and the performance of such measures.

6.8.5.3 Competency

Based on the author's observations and feedback received from numerous operators and asset owners across the oil and gas industry, competency–or more accurately, the lack of competency in the areas of bacterial fundamentals and the associated control measures–has been the main culprit in widespread and often out-of-control bacterial contaminations and MIC issues across the assets concerned.

The competency integrity management measure has four components:

1. Education
2. Experience
3. Training
4. Supervision (or mentoring)

In an ideal situation, all four components are present proportionally. However, under normal circumstances the responsible integrity manager or team leader must ensure that when components 1–3 are lacking or inadequately present, then at least component 4 is fully and thoroughly present and enforced to compensate for the existing deficiencies in other competency areas until they are improved and have reached an acceptable level.

Timely and proper training–which results in more competent personnel–is the ultimate investment and one of the best ways to make integrity cost savings through enhanced (integrity) decision-making and the ensuing implementation. Likewise, for assets where MIC has been one of the main integrity threats, ensuring that the asset corrosion engineer is fully conversant with bacterial basics and the pertaining mitigation options can save significant resources over time through proper and efficient bacterial and MIC mitigation from day one (of the operation).

6.8.5.4 Communication

The lack of, or inadequate, communication has been regarded as the most predominant root cause of corrosion failures and integrity issues across the three sectors of the oil and gas industry. This finding is based both on the author's personal working experience in various countries, and on numerous corrosion and integrity case studies presented or raised by a significant portion of individuals attending the author's asset corrosion management training courses over the years.

Many costly corrosion and integrity failures have been encountered because of the following:

- Failure to hold periodic corrosion meetings attended by the pertinent disciplines (e.g., corrosion, inspection, pipeline, maintenance, operation, and process engineers or representatives)
- Poor and inadequate reporting (e.g., written [hard copy], online, oral)
- Poor, inefficient, and/or impractical reporting formats and templates used (e.g., using very small font sizes and diagram sizes to save paper consumption)

On the contrary, slight enhancements in the *communication* integrity management measure can create significant asset integrity management improvements—at virtually no extra cost. Enhanced communication among the asset integrity team members means better-informed individuals who can see and identify the existing corrosion and integrity issues more clearly and transparently. Consequently, they can devise more efficient, up-to-date, practical, and relevant integrity solutions, and rectification activities.

Throughout the asset integrity management process, different types of data and information are needed or must be created, analyzed, and then acted upon. Simultaneously, many lessons are learned and must be communicated to pertinent personnel or when a situation arises where a collective decision-making is required. This communication process also must occur when a particular discipline has identified a new situation, parameter, or condition with potential repercussions for other disciplines. In all such instances, a clear and effective communication can help to

- Improve or rectify a situation (from the integrity or corrosion viewpoint)
- Get something done
- Get an important message across or notify somebody of an important event, change, or impending incident

When implementing the corrosion management process, regardless of the implementation option taken, the *communication* integrity management measure should always be included within the AIMS. In general, the communication integrity management measure is an indispensable requirement throughout the asset lifecycle, spanning the following asset stages:

- Design
- Construction
- Transportation
- Installation
- Hydrotesting
- Commissioning and operation
- Decommissioning

Total lack of, or inadequate, communication is often one of the many culprits leading into a corrosion issue in general, and MIC in particular. The following are examples of poor communication:

- The operations and process teams or disciplines do not communicate relevant changes–which have detrimental corrosion consequences such as increase in the produced water volumes which also can increase the likelihood of MIC–to the asset corrosion engineer.
- Inspection findings are not communicated to the operation and maintenance teams.
- Critical corrosion topics, parameters, and ranges (such as bacterial population numbers per unit surface or volume) are not communicated to the operation or process teams. That is, the operation team does not learn about the critical integrity condition of a piece of equipment or an area of the asset until it fails, resulting in a leak, explosion, contamination, incident, etc.
- Ordered material or item is different from the purchased material (e.g., software, chemicals, and alloys).
- Relevant integrity data are not communicated to the asset corrosion engineer (such as residual chloride concentration and post-bпромociding bacterial numbers) to enable the engineer to analyze the raw data and make crucial conclusions or determine critical conditions.
- The asset corrosion engineers do not communicate integrity and MIC-related findings to other disciplines.

6.8.5.5 Asset Corrosion Management Strategy (or The Asset Corrosion Control Document)

For many assets, corrosion is a primary integrity threat. To minimize this threat, various inspection, corrosion monitoring, fluid sampling, and chemical injection activities must be regularly carried out–along with the design and materials selection integrity management measures, which have already been in place.

Based on the author's experience, both the effectiveness and efficiency of such activities can be significantly improved when both adequate and up-to-date data, information, and drawings are available. All such data, information, and drawings can be incorporated to produce a very useful and practical corrosion-related and asset-based document–often referred to as the Asset Corrosion Management Strategy (Asset CMS) document, or the Asset Corrosion Control Document (Asset CCD).

An Asset CMS document is a suite of asset-based information and instructions in the form of various corrosion-related databases, drawings, tables, etc. An Asset CMS document can help the asset corrosion engineers to significantly improve their understanding of the asset (i.e., process, operations, design, materials, risk assessment, susceptible areas, etc.) and enable them to make better corrosion engineering and corrosion management judgments and decisions. More specifically, this document can help them to both devise and implement enhanced inspection, corrosion monitoring, fluid sampling, and chemical injection activities. Hence, an Asset CMS can be an indispensable tool to the asset corrosion engineers and the asset integrity management team.

In general, an Asset CMS document should be composed of the following three main sections.

- Section One: Contains all system-based data, information, and drawings:
  - Type of materials or alloys used
  - Fluid composition
  - Existing and expected deterioration mechanisms
  - Existing corrosion mitigation systems and measures
  - Applicable inspection methods
  - A brief integrity history or background of the system
  - Number of corrosion loops, their associated risk ranking, and inspection intervals
  - System process flow diagram with superimposed corrosion loops
- Section Two: Contains various integrity-related data and information:
  - List of all the corrosion rate monitoring systems
  - List of all the various chemicals injected or added
  - List of all locations used for sampling
- Section Three: Contains other pertinent strategies, policies, and any other relevant integrity management tools:
  - MIC mitigation strategy (which can be a combination of various chemical and physical treatments (such as continuous chlorination, regular pipeline cleaning pigging, water removal, and periodic corrosion rate monitoring and fluid sampling)
  - Asset corrosion control matrices
  - Corrosion key performance indicator system

FIGURE 6.1 An example of the types of data that can be summarized within Section One of an Asset CMS.

Produced Oil			31
Piping Material	**Observed Corrosion Mechanism(s)**	**Predicted Corrosion Mechanisms**	**Control Measures**
Carbon Steel	CO_2 Corrosion, Erosion-Corrosion	MIC, CUI, Dead leg Corrosion	Corrosion Inhibitor Injection

Corrosion Rate Monitoring	Two weight loss coupons at the outlet of booster pumps
Fluid Sampling	One sampling location downstream of LP Separator
Susceptible Features	Dead legs, Straights, Reducers, and Welds
Recommended Inspection Techniques	Ultrasonic testing (UT) and General Visual Inspection (GVI)
Representative Fluid(s)	Oil, Condensate, Gas, and Water

Failure Risk Assessment								
Corrosion Loop	**Threat Level Assessment**				**Risk Code**		**Inspection Interval (years)**	
	HAZ	**OPER**	**INT**	**EXT**	**INT**	**EXT**	**INT**	**EXT**
LOP1	H	M	H	L	1	4	4	7
LOP2	H	M	H	L	1	4	4	7

Inspection Details
Carry out a GVI of the whole system and inspect the relevant anomalies. Carry out UT wall thickness inspection of all the selected features on relevant system piping isometrics.

Comments
Carbon steel piping immediately downstream of the HP and LP separators has been more susceptible to corrosion than any other piping sections in this system. The piping section downstream of the LP separator in Train 2 has almost used up its corrosion allowance and has to be scheduled for replacement in first quarter of next year. This year's microbial survey revealed that there is bacterial contamination downstream of the LP separator in Train 2.

6.8.5.6 Procedures

A well-written procedure (document) can significantly improve the comprehension of an (integrity-related) activity or process and facilitate its correct and effective implementation. Such implementation can lead to such advantageous outcomes as lower corrosion rates, fewer corrosion issues, and an improved chemical treatment. Thus, such a procedure document can positively contribute to the overall AIMS. Simultaneously, the lack of such procedures can lead to more severe corrosion problems over time. The following are just two examples.

1. The Pipeline Hydrotesting Procedure: Pipelines are hydrotested prior to commissioning, and an erroneous or even a nonexisting pipeline hydrotesting procedure can inflict significant integrity damage (such as MIC) on a pipeline before it enters the operation phase. Therefore, it is important that a procedure is produced for pipeline hydrotesting, clearly explaining the hydrotest fluid requirements in terms of composition, oxygen content, solid content, pH, chloride content, maximum hydrotest duration, etc.
2. Liquid Sampling Procedure: To ensure that the bacteria and MIC threats are fully controlled and minimized at all times, the asset corrosion engineer should create a liquid sampling procedure. Such a procedure should clearly specify the timing and location of the periodic sampling activities along with all its requirements such as using clean disposable containers each time liquid samples are collected. Proper liquid sampling (and its subsequent analysis) can produce early warnings about the advent of bacterial activity and growth within the asset long before such bacterial contamination has turned into an acute MIC problem.

6.8.5.7 Data Types and Processing (or Data Management)

To carry out successful asset integrity management in general and asset corrosion management in particular, one would rely on and use various data types as their input. Such inputs are then processed (i.e., analyzed or converted) to produce the desired or intended outputs. The inputs and outputs are then used to create an integrity image or picture of the asset where everything can be clearly observed, and in as much detail as possible.

One of the most prevalent issues observed regarding numerous assets and their AIMS is often limiting the (input) data types to inspection data only–particularly when carrying out the integrity review process or the failure risk assessment process. Such an erroneous approach would compel the asset corrosion engineer to view the existing integrity picture or the integrity issues from a rather tight and limited angle. In other words, in such cases, the engineer is only able to view a relatively small portion of the overall integrity picture associated with the asset or any specific piece of equipment.

For any oil and gas asset type, there is a multitude of data types and each type acts as a small piece of a puzzle. One can only see the whole and complete (integrity) picture when all its pieces have been put together. Thus, the more data types used, the more complete the picture looks, and the better one can observe, determine, quantify, and/or identify the existing integrity issues. Consequently, better rectification or integrity solutions can be devised, planned, and then implemented. For example, the data types associated with bacteria and MIC include

- Risk-based inspection data (often collected or created as wall thickness loss data initially)
- Corrosion rate monitoring data
- Liquid sampling data (including bacterial numbers, concentration of various chemical species, water cuts, and residual concentration data)
- Leak data (i.e., location, orientation, and date of the previous leaks including MIC ones)
- Historical chlorination and biocide treatment data
- Historical pipeline cleaning pigging data and the associated pig trash analysis reports
- Historical operation data (number and length of past operation shutdowns)

Any existing shortcoming with the MIC data types, their extent, reliability, and processing can have adverse repercussions on the (asset integrity) picture that must be created. Consequently, the less complete and clear the picture is, the harder it becomes to clearly observe, identify, quantify, and finally rectify the existing (MIC) issues, problems, and shortcomings.

Therefore, data types and their (proper) processing have a significant and crucial influence on how one carries out asset integrity management. Figure 6.2 illustrates the overall data management process.

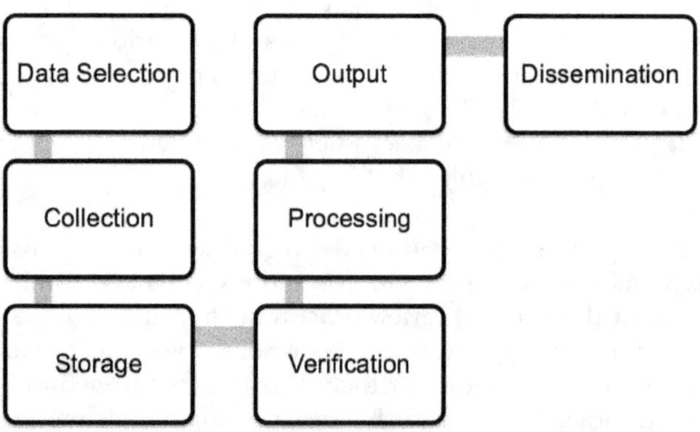

FIGURE 6.2 The overall data processing (or management) process.

Last but not least, to create a better picture of the integrity condition (including the MIC condition) of an asset and its various equipment, as many data types as possible should be used. Figure 6.3 illustrates such data types.

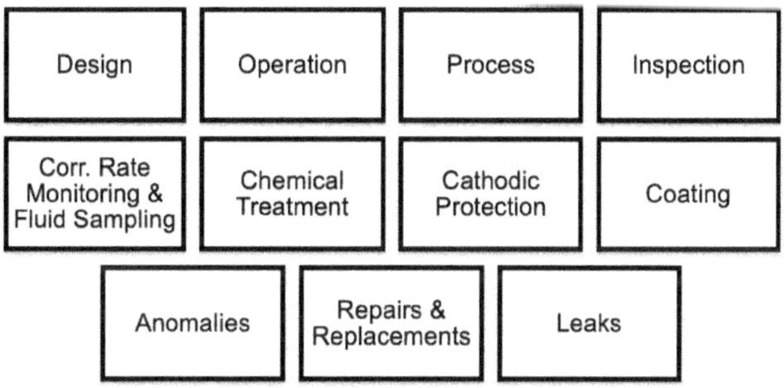

FIGURE 6.3 **Recommended data types for enhanced integrity review process.**

The ultimate or ideal objective of data management should be making all the pertinent and required (input) data digital. Such data digitization would facilitate future data trending, analysis, and the eventual decision-making process when artificial intelligence takes over such activities. Some oil and gas operators are already contemplating such a move.

6.8.5.8 Leak Register

Creating and maintaining a leak register can further facilitate failure data analysis and trending, leading to enhanced corrosion failure preemption capability. Another critical aim of having and maintaining a leak register is to use the relevant leak data in a few different ways:

- Identifying the associated deterioration mechanism(s)
- Determining the shortcomings associated with the existing (corrosion) mitigation measures
- Identifying inspection-related shortcomings that prevented detecting the failure before it occurred

Therefore, for any leak register (database) to be able to perform the aforementioned functions, at a minimum it must have the following items:

- Date (when the leak was detected)
- Material (the type of metal or alloy that leaked)
- Mode of failure (pinhole, crack, or rupture)
- Orientation (or "clock position" of the leak)

- Location (with respect to the immediate upstream equipment or any upstream devices that cause any operational or process changes farther downstream)
- Vicinity to the nearest weld
- Type of equipment suffering from the leak (e.g., piping, pressure vessel, weld, tank)
- Design life vs. actual life
- The last inspection date
- Deterioration mechanism(s) contributing to the leak
- Estimated deterioration rate at the leak site
- Internal or external environment causing the leak
- Existing mitigation methods (which might not have performed well, leading to the leak)

Maintaining and regularly updating the leak register can further facilitate bacterial activity and MIC mitigation. This is because the leak register data would gradually provide a better and improved integrity picture about the locations and nature of each MIC leak; thus, the operator or asset owner can increasingly make more informed decisions regarding how to control and resolve any existing MIC issues and problems.

6.9 Summary

Depending on the type, size, and nature of the operation of an asset, as well as the type, extent, and severity of the associated bacterial and MIC issues, an operator can implement one or a combination of the following activities to combat bacterial activity and control MIC within their asset:

- Chlorination
- Biocide treatment
- Water removal (or product dehydration)
- Pipeline cleaning pigging
- Nitrate treatment

Simultaneously, proper and timely application of corrosion management principles can significantly enhance the effectiveness of the associated bacterial and MIC mitigation process.

7 Shortcomings Associated with the Bacterial and MIC Mitigation Process

7.1 Introduction

To effectively control any bacterial contamination and the associated MIC integrity threat, the *bacterial and MIC mitigation process* must be properly implemented. More precisely, each of the three stages composing the mitigation process must be carried out in a timely, organized, proper, and faultless manner. Any potential shortcomings associated with any of these three stages would undoubtedly detrimentally affect the whole mitigation process and its overall effectiveness and outcome.

This chapter lists various shortcomings encountered by the author over a period of 20 years during site visits, integrity reviews, failure investigations, and asset integrity management system (AIMS) audits. These shortcomings pertain to the three stages of the bacterial and MIC mitigation process. The main objective of this chapter is to share valuable international experience and learn from the mistakes and blunders of operators, organizations, and individuals in the area of bacterial and MIC mitigation. The ultimate goal is to ensure that such mistakes and erroneous conduct will never be repeated elsewhere.

7.2 Categorizing the Observed Shortcomings

Almost all of the shortcomings listed in this chapter were identified or encountered during site visits, integrity reviews, failure investigations, and asset integrity management system (AIMS) audits carried out by the author over the past 20 years in various parts of the world.

Most of the shortcomings listed in this chapter are associated with either seawater injection systems or various types of inlet water systems, or those systems located further downstream of such systems. Such inlet water systems filtered raw water and then chemically treated it prior to its being used as cooling water, fire water, sand wash water (for washing and removing the sand accumulated in pressure vessels), or any other (mainly utility) applications within the assets concerned.

To be able to more clearly list and describe the associated shortcomings, they have been classified into three different categories. This categorization system is identical to the way the activities associated with the bacterial and MIC mitigation process were categorized:

- Bacterial and MIC monitoring stage shortcomings
- Bacterial and MIC assessment stage shortcomings
- Bacterial and MIC control stage shortcomings

Similarly, based on the author's observations when dealing with and studying such shortcomings, the root cause of the majority, if not all of them, was total or partial lack of competency pertaining to bacterial and MIC fundamentals (Chapters 4–6) among the relevant personnel and their managers.

Therefore, to avoid such shortcomings in future, the relevant personnel and managers should improve their bacterial and MIC competency sooner rather than later via self-study, attending training courses, or other means.

7.3 Important Reminder to the Reader of this Chapter

The author has done his best to describe each individual shortcoming as clearly as possible to ensure it is fully comprehended by the interested reader. However, thereafter, the onus is on the reader to

- Ensure that the relevant shortcoming and its root cause(s) are completely understood, and then to
- Ensure that the same mistake (leading to a similar shortcoming) in their asset and operation will not be repeated

Learning from other operators' mistakes (in the areas of bacteria and MIC) and avoiding them can induce significant savings in various resources and simultaneously increase asset uptime and longevity. This is the main objective of this chapter (and the author).

7.4 Bacterial and MIC Monitoring Stage Shortcomings

The shortcomings associated with the bacterial and MIC monitoring stage are further divided into the following two groups:

- Bacterial monitoring activities
- MIC monitoring activities

Sections 7.4.1 and 7.4.2 cover the above two shortcoming categories, respectively.

7.4.1 Bacterial Monitoring Activities Shortcomings

The observed shortcomings associated with the bacterial monitoring activities include:

- Not taking any samples at all (both liquid and solid or planktonic, and sessile [or biofilm])
- Only taking liquid (or planktonic) samples
- Using the same bottle or container for taking liquid samples from various locations
- Not fully filling up the sample bottles to drive out air and then not using an airtight cap to seal it properly
- Leaving sample bottles uncapped in the environment for long periods of time where their content might get contaminated
- Leaving sample bottles under the sunlight and exposed to extreme heat conditions (in hot climates) for long periods of time
- Taking multiple samples from one single point at one particular time but labeling them in such a way as if they have been taken from various locations across the asset at different time intervals
- Taking samples from (process) locations where no water (and therefore no bacteria) is present
- Taking samples irrespective of the biocide treatment time intervals (samples should be taken both before and after each biocide treatment)
- Taking samples from wrong locations (e.g., where bacteria would never grow)
- Taking samples from wrong orientations (e.g., taking liquid samples from the 12 o'clock position instead of the 6 o'clock position
- Taking solid (or biofilm) samples, but not properly preserving them when the intended assessment methods require sample preservation

7.4.2 MIC Monitoring Activities Shortcomings

The observed shortcomings associated with the MIC monitoring activities include the following:

- Not carrying out risk-based inspections. The majority of inspections being carried out are time-based or in some cases consequence-based
- Carrying out spot (wall thickness) inspections instead of band (wall thickness inspections)
- Often inspecting the top or the upper half of equipment (such as piping and pipelines) instead of the lower half or the 6 o'clock position
- Not having a dead leg corrosion policy, which often specifies more rigorous inspection and/or chemical treatment of such areas within an asset
- Installing the corrosion monitoring equipment (such as electrochemical probes and weight loss coupons) at the wrong locations (from a bacterial contamination and MIC viewpoint)
- Installing the corrosion monitoring equipment at the wrong orientation (e.g., the 12 o'clock position instead of the 6 o'clock position)
- Not taking biofilm samples or bacterial slimes from the retrieved probe or coupon holders
- Not keeping the biofilm samples taken in airtight containers
- Not using the produced pig trash for lab analysis and assessment (at the end of a pigging activity)
- Not removing the weight loss coupons in time so they are totally consumed; thus, it is impossible to calculate or estimate the corrosion rate, and to take bacterial biofilm samples from the coupon surface

7.5 Bacterial and MIC Assessment Stage Shortcomings

The shortcomings associated with the bacterial and MIC assessment stage are further divided into the following two groups:

- Bacterial assessment activities
- MIC assessment activities

Sections 7.5.1 and 7.5.2 cover the above two shortcoming categories, respectively.

7.5.1 Bacterial Assessment Activities Shortcomings

Following are the observed shortcomings associated with the bacterial assessment activities:

- Only using the traditional or culture-based methods to count bacteria
- Using out-of-date, uncalibrated, unclean, imprecise, and unreliable lab equipment
- Using the wrong culture type (or medium or nutrition) for growing a specific type of bacteria
- Incompetent lab staff or technicians carrying out the assessments
- Not following the correct or required procedure when conducting the serial dilution or the most probable number methods

7.5.2 MIC Assessment Activities Shortcomings

Following are the observed shortcomings associated with the MIC assessment activities:

- Erroneously attributing corrosion rates due to abiotic corrosion mechanisms (such as under deposit corrosion) to MIC
- Erroneously reporting no or negligible MIC deterioration rates due to the corrosion rate monitoring devices (probes or coupons or both) being fitted at the 12 o'clock position instead of the 6 o'clock position
- Erroneously reporting no or negligible MIC deterioration rates because the wall thickness loss inspections carried out were not risk-based, and the wrong locations (where MIC did not exist) were inspected instead

7.6 Bacterial and MIC Control Stage Shortcomings

The observed shortcomings associated with the bacterial and MIC control stage are discussed in the following sections, based on the nature and type of the bacterial and MIC control activities discussed in Chapter 6.

7.6.1 Chlorination Shortcomings

Following are the observed shortcomings associated with the chlorination activity:

- Carrying out no chlorination at all
- Commissioning the chlorination equipment months *after* commissioning the associated inlet water system or the seawater system instead of commissioning them simultaneously
- Carrying out batch or intermittent chlorination instead of a continuous one
- Carrying out chlorination at alkaline pH values—where the oxidizing power of chlorine, and hence its killing power, significantly decreases
- Adding chlorine at a location where it does not mix adequately with inlet water
- Adding chlorine at an inadequate concentration, leading to inadequate residual chlorine levels in the system
- Adding chlorine, but with inadequate residence time, i.e., the added chlorine does not have enough time to achieve a complete or more efficient (planktonic) bacterial kill (for chlorination a residence time of up to 30 min might be required to achieve a complete kill)
- Adding chlorine downstream of the oxygen scavenger injection point. This practice renders the injected chlorine useless, because it reacts and is deactivated by the present oxygen scavenger chemical
- Carrying out continuous chlorination, but for only a small portion of the inlet water. This often happens in facilities that possess huge pools for holding water along with multiple chlorination points, of which only one or a few func-

tion properly on a continuous basis and the majority (of the injection points) are (somehow) out of service

- Carrying out chlorination and biocide treatment at the same location where their injection points are located at close proximity with each other
- Carrying out chlorination at certain locations to kill sessile bacteria and not planktonic bacteria (chlorine is not effective against sessile bacteria or biofilms)

7.6.2 Biocide Treatment Shortcomings

Following are observed shortcomings associated with biocide treatment activity:

- Injecting the biocide chemical at such a low concentration that it does not kill the existing bacteria
- Injecting a biocide chemical that is only capable of killing planktonic bacteria and is ineffective against sessile bacteria and biofilms
- Injecting a biocide chemical at a wrong frequency rendering it inefficient against the existing bacteria
- Injecting biocide at such operating conditions which can deactivate or decompose it; such as injecting at operating temperatures, which are higher than the maximum operating temperature specified for that particular biocide
- Injecting a biocide chemical that lacks chemical compatibility with other injected chemicals and is thus inactivated upon injection
- Injecting biocide at the wrong location (e.g., much further downstream of the required location, thus leaving the systems and equipment upstream of the injection point unprotected against the bacteria and MIC)
- Injecting a biocide chemical for such a long time that the target bacterial populations gradually become immune against that biocide
- Injecting biocide at the wrong pH value or the wrong pH range (some biocide chemicals perform best at certain pH values, or pH ranges based on their datasheets)
- Injecting biocide chemicals that lack surface active agents and thus cannot penetrate into biofilms to kill the sessile bacteria underneath the biofilm
- Injecting the biocide chemical upstream of the oxygen scavenger chemical instead of injecting it downstream of that chemical

7.6.3 Water Removal Shortcomings

The observed shortcomings associated with the water removal activity include

- Not completely removing water from the process system concerned (e.g., incomplete gas dehydration); thus, the bacterial activity and MIC threat remains
- Not completely draining water that accumulated at the bottom of various equipment (e.g., the water present at the bottom of product tanks such as diesel)

7.6.4 Pipeline Cleaning Pigging Shortcomings

The observed shortcomings associated with pipeline cleaning pigging include

- Using the wrong type of pig that is incapable of properly removing the water and/or sufficiently scrape the biofilm(s)
- Using old pigs with deflected or damaged parts rendering them incapable of proper and adequate water (and biofilm) removal
- Using pigs already contaminated with bacteria and biofilm debris in clean pipelines, thus contaminating them as well
- Carrying out pigging operations infrequently, thus letting bacteria and the associated biofilms to settle at the bottom of pipelines and causing significant damage

7.6.5 Nitrate Treatment Shortcomings

Following are the shortcomings associated with nitrate treatment:

- Injecting nitrate as a batch treatment instead of a continuous one
- Not carrying out corrosion rate monitoring and wall thickness inspections downstream of the nitrate injection point (because some nitrate-reducing bacteria themselves cause MIC and thus frequent corrosion rate monitoring and wall thickness inspections are required to identify such possibility as soon as possible)

7.6.6 Corrosion Management Applications Shortcomings

Following are the observed shortcomings associated with corrosion management application activities:

- Not specifying corrosion control matrices (also known as "plant integrity windows" or "operating integrity windows")
- Not specifying and reporting corrosion key performance indicators
- Not processing various data (such as wall thickness losses, corrosion rate monitoring, liquid sampling, bacterial populations, chemical injection doses, and residual chlorine levels), thus not being able to determine various corrosion parameters (such as deterioration mechanisms, deterioration rates, and locations of ongoing corrosion and wall thickness loss)
- Not having procedures or not having up-to-date and relevant procedures (such as the liquid sampling procedure)
- Not having adequate competencies in general, or in the area of bacteria and MIC in particular (this item is often associated with the asset corrosion engineer)
- Not adequately sharing and disseminating relevant data (such as operation data, process data, inspection data, sampling data, chemical treatment data, and [MIC] leak data)

7.7 Summary

To ensure the implementation of an effective *bacterial and MIC mitigation process*, the personnel concerned must make sure that each of the three stages that compose this process are carried out properly.

Operators can ensure that they implement their bacterial and MIC mitigation process more efficiently by simultaneously learning from other operators' mistakes and making every effort to avoid them.

8 Conclusions and Recommendations

8.1 Conclusions

There are a multitude of integrity threats associated with the upstream oil and gas sector assets, and corrosion is often a major one. Among various corrosion deterioration mechanisms encountered in the upstream sector, MIC is considered as one of the most prevalent and insidious ones. It not only damages the upstream assets, but also–if not adequately controlled and mitigated–has the potential to damage the associated reservoirs.

Thus, every effort must be made to efficiently and satisfactorily control both bacterial activity and the associated MIC threat. However, combating bacterial contamination and MIC has been very tricky and many operators have not been capable of creating and/or implementing an efficient *bacterial and MIC mitigation process*. Any such process should be composed of the following three stages:

- Bacterial and MIC monitoring stage
- Bacterial and MIC assessment stage
- Bacterial and MIC control stage

Based on the author's personal experience over the past 20 years, the main culprit–and root cause–behind the frequent and often severe MIC issues encountered in the assets concerned is the lack of adequate competency with bacterial nutrition and growth conditions with respect to the above three stages. Therefore, to resolve this issue and improve the situation, asset owners and operators who have to deal with bacterial contamination and MIC issues must ensure that the pertinent personnel and managers are fully knowledgeable on *bacterial and MIC basics*.

8.2 Recommendations

This section lists the author's recommendations and are categorized into four groups:

- Bacterial and MIC monitoring stage recommendations
- Bacterial and MIC assessment stage recommendations
- Bacterial and MIC control stage recommendations
- General recommendations

The most crucial recommendation, however, is to ensure the pertinent personnel are adequately trained and competent in bacterial and MIC basics.

8.2.1 Bacterial and MIC Monitoring Stage Recommendations

- Take both liquid and solid (i.e., sessile or biofilm) samples.
- Create a procedure for sampling and include all relevant details such as locations, frequency, and timing of the biocide treatments.
- Use disposable equipment for sampling.
- Always include a baseline sample (i.e., sample taken from a location upstream of the first chemical treatment, chlorination, or biociding) in the sampling procedure to enable comparisons with it to treated water qualities (e.g., residual chemical, concentration of various chemical species, and bacterial populations).
- Completely fill up sample bottles and use airtight caps to ensure no sample oxidation or contamination takes place during storage and transport.
- Ensure that the applicable weight loss coupons and/or the electrochemical probes are fitted at the correct locations and orientation (i.e., the 6 o'clock position).

Sections 2 and 3 of Reference 1 also provide invaluable advice regarding proper sampling.

8.2.2 Bacterial and MIC Assessment Stage Recommendations

- Use both traditional (i.e., culture-based) and molecular microbiological methods simultaneously to determine bacterial types and numbers.
- Only use labs that have all the necessary certificates pertaining to equipment, procedures, and personnel's competency.
- Compare corrosion rates calculated from wall thickness loss inspections with those from weight loss coupons and electrochemical probes. If such rates are very different, try to find out which one more accurately represents the ongoing MIC rates.

8.2.3 Bacterial and MIC Control Stage Recommendations

- Apply chlorination on a continuous basis.
- Ensure that the injected chlorine is fully mixed with the inlet water phase.
- Allow adequate residence time to ensure that more planktonic bacteria have been killed by the injected chlorine.
- If multiple chlorination points are used, always make sure that all points are functional and working.
- During chlorination treatment, avoid adding alkaline chemicals to make pH values more alkaline for the inlet water.
- Do not add chlorine and a biocide chemical at close proximity and/or simultaneously because chlorine requires a continuous treatment and biocide chemicals require a batch treatment.
- If an optimum pH value has been specified for the selected biocide chemical, ensure that it is injected at that optimum pH value or range.
- Ensure that the chemical being purchased as the biocide is not corrosive to carbon steel or other existing alloys.
- Ensure that the chemical being purchased as the biocide is capable of killing *sessile* bacteria and not only planktonic bacteria. Simultaneously ensure that is has adequate surface active agents to penetrate the biofilms to kill the sessile bacteria inhabiting at the biofilm and metal interface.
- Use solid biocide pills or rods for new clean (wet) storage systems that have not shown any sign of MIC yet.
- When specifying or creating a biocide treatment procedure (detailing frequency and concentrations) ensure that the activity value of the biocide chemical is known and taken into consideration when specifying the minimum required concentration or injection rate.

8.2.4 General Recommendations

- Ensure that relevant personnel are adequately competent in the areas of bacterial and MIC basics.
- Create a corrosion control matrices document (if it does not exist already) or regularly update and revise it (if it already exists).
- Ensure that corrosion key performance indicators are created and used, and that they include relevant bacterial and MIC items wherever applicable.
- If necessary, create a bacterial and MIC mitigation policy that details all the required monitoring, assessment, and control activities.
- When carrying out data management, include and use various data types such as
  - Wall thickness inspection data
  - Corrosion rate monitoring data
  - Sampling data
  - Operation data
  - Process data
  - Chemical treatment data
  - Pig trash analysis data (for pipelines)
  - Leak data
  - Anomaly data

References

1. NACE TM0194, "Field Monitoring of Bacterial Growth in Oil and Gas Systems" (Houston, TX: NACE International, 2014).
2. Morshed, A., *An Introduction to Asset Corrosion Management in the Oil and Gas Industry*, 3rd Ed. (Houston, TX: NACE International, 2020).
3. Battey, N., and M. Fellows, eds., *30-Second Biology* (London, United Kingdom: Ivy Press, 2016).
4. Skovhus, T.L., and C. White, eds., *Oilfield Microbiology* (Boca Raton, FL: CRC Press, 2019): p. 86.
5. Skovhus, T.L., D. Enning, and J.S. Lee, *Microbiologically Influenced Corrosion in the Upstream Oil and Gas Industry* (Boca Raton, FL: CRC Press, 2017): p. ix.
6. NACE TM0212, "Detection, Testing and Evaluation of Microbiologically Influenced Corrosion on Internal Surfaces of Pipelines" (Houston, TX: NACE International, 2012).
7. Larsen, J., "Downhole Nitrate Applications to Control Sulfate Reducing Bacteria Activity and Reservoir Souring," CORROSION 2002, paper no. 02025 (Houston, TX: NACE International, 2002).

MIC Case Studies

Introduction

This appendix contains case studies pertaining to MIC issues associated with upstream, midstream, and downstream sectors of the oil and gas industry. The case studies pertain to actual integrity management projects, audits, or asset integrity reviews carried out by the author. The assets covered include offshore production platforms, subsea and overland pipelines, onshore terminals, and downstream processing facilities.

The case studies detailed here follow the same structure with these components:

Overview–A brief summary of the asset and the encountered bacterial and MIC situation.

Observation–The main finding(s) of the author during his visit or integrity review.

Culprit–The main shortcoming(s) and/or noncompliance(s) identified.

Outcome–The most significant repercussion(s) of the identified shortcoming(s).

Recommendations–The main rectifying actions suggested by the author (whether they were implemented by the asset owner or not).

Note: To protect the location and the true identity of the asset owners/operators, and the contractors, the physical description (i.e., the location, number of trains, age, or any other physical characteristics) of the assets covered in the following case studies have been slightly modified.

Case Study 1: An Onshore Seawater Treatment System

Overview: An onshore seawater treatment system treats around 10 million barrels of seawater on a daily basis. The plant and the pipeline network downstream of it (and extending for more than 500 km) both suffer from severe corrosion and numerous leaks, thus requiring regular (often costly) repair and maintenance shutdowns. The operator requests an integrity review to be carried out.

Observation: During the site visit and the ensuing asset integrity review it became evident that chlorination was inadequate and intermittent, but more importantly biocide treatment had been stopped for a long time. Such poor bacterial control (inadequate chlorination along with no biocide treatment) had resulted in frequent corrosion leaks within the plant and numerous pipeline leaks along the pipeline network downstream of the seawater treatment plant.

Culprit: The penultimate plant manager had reasoned that *because bacteria were too small to be seen with naked eyes, the integrity threat they posed was negligible accordingly* and consequently he had ordered the biocide treatment activity to be completely stopped. Simultaneously, relevant personnel under him were not adequately conversant with bacterial and MIC basics.

Outcome: The plant operated at around 70% of its nominal capacity with significantly increasing repair costs. The extra repair budget requested for one single year was $100 million. Total repair costs were estimated to be significantly higher with more operation shutdowns planned.

Bacteria and MIC Recommendations:

- Improve chlorination by changing it to a continuous treatment at correct concentrations
- Resume biocide treatment
- Enhance sampling activities and create a new sampling procedure for the plant
- Organize a bacteria and MIC training course for the pertinent personnel

Case Study 2: An Onshore Production Plant

Overview: An onshore production plant produces oil and gas, and uses the nearby river as a source of utility water for cooling purposes. The cooling water system suffers from severe MIC issues requiring frequent operation shutdown and maintenance. The operator requests a site visit and an integrity review to be carried out.

Observation: During the site visit and the ensuing interviews with the pertinent personnel, it was revealed that chlorination was carried out on an intermittent basis–in spite of the fact that the river water was flowing in through the water inlet system on a continuous basis.

The cooling water system, its piping, tanks, and pressure vessels were suffering from heavy bacterial contamination, and MIC requiring frequent operation shutdowns, repairs, and maintenance campaigns.

Culprit: No one at the site had realized that if raw water was flowing in on a continuous basis, then the chlorination activity had to be also continuous and not intermittent. The planktonic bacteria that were entering the plant were partly treated with the intermittently injected chlorine and the surviving ones had initiated severe MIC issues farther downstream and across the cooling water system.

Outcome: Frequent cooling water system shutdowns due to MIC issues detrimentally affected plant production rates and markedly increased repair and maintenance costs.

Bacteria and MIC Recommendations:

- Change chlorination from an intermittent treatment to a continuous one
- Organize a bacteria and MIC training course for the pertinent personnel and managers
- Revise the existing bacterial and MIC mitigation process

Case Study 3: An Offshore Seawater Injection System

Overview: An offshore seawater treatment system in the United Kingdom's North Sea Sector uses both chlorination and biocide treatment to control and mitigate bacteria and MIC. However, wall thickness loss inspections along with the existing corrosion rate monitoring system point to high corrosion deterioration rates downstream of the biocide injection point. The operator requests an asset integrity review to be carried out.

Observation: The integrity review revealed that the biocide chemical being used was only capable of killing *planktonic* bacteria and was incapable of penetrating into the biofilms to kill the *sessile* bacteria entrenched there. Increasing numbers of operation shutdowns (of the seawater injection system) and spiraling repair and maintenance costs were also the consequences of this situation.

Culprit: The purchased biocide chemical was only capable of killing *planktonic* bacteria per the lab report that had certified the chemical. However, the associated personnel and managers did not know that any typical biocide chemical had to be able to kill *sessile* bacteria as well. In general, they did not know the differences between *planktonic* and *sessile* bacteria.

Outcome: Severe bacterial contamination of the seawater injection system along with frequent MIC leaks and operation shutdowns also associated with increasing repair costs.

Bacteria and MIC Recommendations:

- Purchase a new biocide chemical capable of penetration into biofilms and killing sessile bacteria
- Organize a bacteria and MIC training course for the pertinent personnel

Case Study 4: An Onshore Cooling Water System

Overview: An onshore cooling water system is associated with an upstream hydrocarbon production plant. The cooling water system suffers from severe MIC issues with frequent operation shutdowns for the required repair activities. The operator requests a corrosion audit to be carried out.

Observation: During the site visit and the ensuing asset integrity review, it became evident that chlorination and biocide treatment were carried out simultaneously at one location at the entrance of the water inlet system. An alkaline chemical was also added into the system to increase the pH value and make inlet water pH slightly alkaline.

Culprit: The site personnel were under the impression that if they injected chlorine and the biocide chemical together at slightly alkaline conditions, then such chemical treatment would kill all the bacteria within the inlet water system. Their poor understanding of the bacteria and MIC control basics resulted in an ineffective chemical treatment system.

Outcome: Frequent operation shutdowns of the cooling water system detrimentally affected the overall production rate of the plant with increasing repair costs due to the exacerbating MIC issue.

Bacteria and MIC Recommendations:

- Move the biocide injection system farther downstream of the deaeration system
- Inject chlorine at slightly acidic conditions (to improve the efficiency of the chlorination treatment)
- Organize a bacteria and MIC training course for the pertinent personnel

Case Study 5: A Subsea Seawater Transfer Pipeline

Overview: A subsea seawater transfer pipeline is used for transferring treated seawater from an offshore production platform to multiple subsea seawater injection wells to maintain reservoir (production) pressure. An intelligent pig run demonstrates that the pipeline has suffered from severe internal corrosion. The operator requests a pipeline integrity review to be carried out.

Observation: Based on the findings of the intelligent pig run, the pipeline suffered severe MIC at and around the 6 o'clock position. Review of the available chemical treatment data showed that chlorination was intermittent while biocide injection was carried out frequently; however, there was almost no record or data of bacterial numbers pre- and post-biociding.

Culprit: The site personnel were not aware of the significance of continuous chlorination. Furthermore, nobody on site knew whether the biocide that was injected for the past several years was actually killing any bacteria or not. No sampling was carried out pre- and post-biociding so bacterial numbers before and after biociding were unknown. Thus, nobody knew how effective the biocide treatment was.

Outcome: Severe internal corrosion at the bottom of the pipeline markedly increased the likelihood of its replacement in the foreseeable future at a significant cost and well before the termination of its intended design life.

Bacteria and MIC Recommendations:

- Create a procedure for both liquid and biofilm sampling for the associated pipeline and its upstream seawater treatment asset
- Organize a bacteria and MIC training course for the pertinent personnel and managers

Abbreviations and Acronyms Used in the Literature

AIM	Asset Integrity Management
AIMS	Asset Integrity Management System
AMPP	Association for Materials Protection and Performance
APB	Acid Producing Bacteria
APS	Adenosine-5-Phosphosulfate
ATP	Adenosine Triphosphate
CCD	Corrosion Control Document
CCM	Corrosion Control Matrices
CE	Corrosion Engineering
CI	Corrosion Inhibitor
CM	Corrosion Management
CMS	Corrosion Management Strategy
CP	Cathodic Protection
CRA	Corrosion Risk Assessment

CRA	Corrosion Resistant Alloy
CUI	Corrosion Under Insulation
DAPI	4′,6-diamino-2-phenylindole
DNA	Deoxyribonucleic Acid
EAC	Environmentally-Assisted Cracking
EXT	External
FISH	Fluorescence In Situ Hybridization
FRA	Failure Risk Assessment
FW	Fire Water
GHB	General Heterotrophic Bacteria
GVI	General Visual Inspection
HAZ	Hazard
HP	High Pressure
HSE	Health and Safety Executive
HS&E	Health, Safety and Environment
IM	Integrity Management
INT	Internal
IRP	Integrity Review Process
KPIs	Key Performance Indicators
LP	Low Pressure
LR	Lloyd's Register
MAWT	Minimum Allowable Wall Thickness
MCMM	Morshed Corrosion Management Model

MIC	Microbiologically Influenced Corrosion
MMM	Molecular Microbiological Methods
MO	Microorganism
MP	Materials Performance
MPN	Most Probable Number
mpy	mils per year
MS	Microsoft
MSD	Materials Selection Diagram
NACE	National Association of Corrosion Engineers
NGS	Next Generation Sequencing
OPER	Operability
P	Pressure
PFD	Process Flow Diagram
P&IDs	Piping and Instrumentation Diagrams
ppb	Parts per billion
ppm	Parts per million
PREN	Pitting Resistance Equivalent Number
PSN	Production Services Network
PSV	Pressure Safety Valve
PW	Produced Water
qPCR	Quantitative Polymerase Chain Reaction
RBI	Risk-Based Inspection
SRB	Sulfate-Reducing Bacteria

SW	Seawater
T	Temperature
UDC	Under Deposit Corrosion
UK	United Kingdom
UT	Ultrasonic Testing

www.ingramcontent.com/pod-product-compliance
Lightning Source LLC
LaVergne TN
LVHW022005200926
848481LV00007B/2246